Planning for Field Safety

American Geological Institute
Alexandria, Virginia

Cover illustration by Shelly Fischman
Cover design by Julie De Atley
Typeset in Times Roman using Ventura Publisher

Copyright © 1992 by the American Geological Institute

All rights reserved. No part of this book may be reproduced in any manner without the express written consent of the publisher, except in the case of brief excerpts in critical reviews and articles.

Library of Congress catalog card number 89-083681
ISBN 0-913312-93-2

Printed in the United States of America

American Geological Institute
4220 King Street
Alexandria, VA 22302-1507
(703)379-2480

CONTENTS

PREFACE vii

1 PRETRIP PLANNING 1

Checking Out the Weather
Choosing Equipment
   Essential Gear, Clothing, Footgear, Work Gloves, Tools, Rope, Backpacks, Tents, Stoves, Campfire Starters
Medical Preparations
   Immunizations and Medications, Medical History, Personal and Base Camp Medical Kits, Supplementary Medications, Medical Kit Guidelines
Food
Logistics
Evacuation Plan
Personal Preparation
Safety Regulations

2 EQUIPMENT PRECAUTIONS 21

Tools and Equipment
   Tools, Stoves, Gas Lanterns, Firearms
Drilling Rigs and Sites
   Individual Safety Equipment, Hand Tools, Drilling Site

3 SAFETY IN THE FIELD 37

Fieldwork
   Marijuana and Moonshine, Emergency Plans, Transportation, Keeping Track, Protective Gear,

    Risky Conditions, Avoiding Falls, Crossing Streams,
    Quicksand, If You Get Lost
  Food and Water
    Contaminated Food, Contaminated Water
  In Camp
    Campfires
    Wildfires

4 TRANSPORTATION   59

  On and Off the Road, Snow Vehicles, Pack
  Animals, Airplanes, Helicopters, Safety on the
  Water, Motorboats, Canoes and Inflatable Boats,
  Tidal Areas, Oceanic Research Vessels

5 WEATHER CAUTIONS   89

  Cold Weather
    Clothing, Hypothermia, Frostbite, Cold-water
    Immersion, Snow
  Hot Weather
    Dehydration, Heat Illness, Sunburn
  Thunderstorms
    Lightning

6 ANIMALS AND PLANTS   115

  Animals
    Bears, Other Animals
  Snakes
  Insects and Other Pests
    Mosquitoes and Flies; Bees, Wasps, and Ants;
    Spiders; Scorpions; Ticks; Chiggers
  Poisonous Plants

7 **HAZARDS OF SPECIFIC REGIONS
AND SETTINGS** 141

  High Altitude
    Acclimatization, Acute Mountain Sickness,
    Symptoms of Severe AMS
  Alpine Regions
  Tundra
  Tropics
    Hawaii
  High and Low Desert and Semiarid Regions
  Temperate Forested Regions
  Underground Mines and Caves
  Fieldwork in Other Countries
    Preparations, On the Road, Local Conditions
  Urban Environment
  Day Trips
    Before the Trip, In the Field

8 **IN CASE OF EMERGENCY** 167

  Patient Assessment System
    Primary Assessment, Secondary Assessment
  Evacuation Procedures
  Sample Patient Assessment Form

9 **PRECAUTIONS FOR THE GROUP LEADER** 181

10 **SHIPPING ROCK SAMPLES** 185

  Containers, Shipping from Other Countries

  INFORMATION SOURCES 189

  INDEX 193

PREFACE

This book describes the more common hazards and pitfalls fieldworkers encounter, suggests ways to avoid them, and tells what to do if they occur. Its goal is to make fieldworkers more conscious of safety, so that safe fieldwork becomes a matter of careful planning, not uncertain luck. These guidelines for wilderness safety were compiled first for students of geology; however, others, including hikers, backpackers, and fieldworkers in any profession will find useful information here.

Readers should bear in mind that this manual is a broad introduction to field safety. Although it should prove useful to fieldworkers in a wide variety of settings and environments, a book of this scope and size cannot cover all topics in exhaustive detail. Before heading into the backcountry, you should thoroughly investigate the field conditions and potential hazards in your specific area. A reading list at the back of the book will help guide you to more detailed sources of information on topics that pertain to your trip.

All fieldworkers are strongly urged to become certified in basic first aid and cardiopulmonary resuscitation. Such training will provide the hands-on experience necessary to respond to field emergencies quickly, confidently, and effectively—and can help prevent a minor crisis from becoming a major disaster. Contact your local American Red Cross chapter or health organizations for

PLANNING FOR FIELD SAFETY

information on training courses in your area. In addition, the National Association for Search and Rescue (NASAR) offers a series of medical training programs specifically designed for outdoor group leaders and others who may need to provide emergency care in remote areas. For further information, contact NASAR Wilderness Medicine Programs, Wilderness Medical Associates, RFD 2 Box 890, Bryant Pond, ME 04219 (207/665-2707).

No matter what type of fieldwork you do or where you do it, remember that a safety-conscious attitude is the most effective defense against danger. Be alert. Know your limitations and don't exceed them. Common sense, sound judgment, and an eye out for hazards will guide you safely through most field activities.

* * *

This guide was compiled by Virginia Sisson and Karen Kleinspehn while graduate students in the Department of Geological and Geophysical Sciences at Princeton University, and was later expanded and edited by Gerald O'Reilly for the American Geological Institute (AGI). It is the product of the thoughts and experiences of many people. The compilers, their present employers (University of Minnesota and Rice University), Princeton University, and AGI do not assume any liability for personal injury, including death or damage to property or financial loss that may be sustained by readers or users of this guide. In addition, the inclusion

PREFACE

of brand names of products does not mean that the compilers, Princeton University, or AGI endorses those products.

The guide was inspired by a University of Massachusetts manual by Stearns A. Morse, *Draft Safety Policy for Field Workers in Remote Areas,* and by a U.S. Geological Survey publication by David Brew and others, *Guidelines for Safe Geologic Fieldwork in Alaska;* various parts of these works have been incorporated into this guide.

Much of the health and first aid information presented here originally came from the Princeton University Health Services, Health Education Office. That information has been updated based on material from the National Association for Search and Rescue and other sources and reviewed by Dr. Peter Goth, Program Director, NASAR Wilderness Medicine.

Our thanks go to Bruce Douglas, Indiana University, Bloomington; Noel Potter Jr., Dickinson College; Joseph Nadeau, Rider College; Marvin Kauffman, National Science Foundation; David Howell, U.S. Geological Survey, Menlo Park, California; Christopher Mathewson and David White, Texas A&M University, College Station; Virginia Sisson, Rice University; and Karen Kleinspehn, University of Minnesota, for their review of the manuscript and suggestions for improvement.

1

PRETRIP PLANNING

Safe fieldwork is no accident. It requires careful planning of all aspects of your trip—equipment, food, clothing, logistics. This chapter will help guide your planning.

Checking Out the Weather

Your choice of tools, clothing, and other equipment is largely determined by the environment you expect to encounter and the work you'll be doing there; the harsher the environment, the more critical these choices become. It is vital, therefore, to be aware of potential weather extremes and prepare accordingly.

Before heading into an unfamiliar area, find out what sort of weather you can expect by calling local weather bureaus, consulting area residents, and talking to fieldworkers who have spent time in the area during the season you'll be working there. In the United States, the National Oceanic and Atmospheric Administration (NOAA) operates a network of radio stations that broadcast reliable weather information around the clock, 365 days a year. About 90% of the country is now within

PLANNING FOR FIELD SAFETY

range of one of these stations (although to hear one you'll need a small radio capable of receiving the very high frequency "weather band"). In areas not covered by weather broadcasts, a small altimeter can be effective to monitor rapid drastic drops in air pressure and the onset of storm systems. For a complete list of NOAA weather stations, write NOAA, National Weather Service, 8060 13th Street, Silver Spring, MD 20910; Attention: W112. Another source of U.S. weather information is a publication series titled *Comparative Climatic Data*, available from the U.S. National Climatic Center, Federal Building, Asheville, NC 28801-2696. The series will give you information, usually by month, such as average snowfall, average rainfall, and temperature extremes, for a given region.

For Canada, temperature and precipitation tables for each province are available from the Assistant Deputy Minister, Atmospheric Environment Service, 4905 Dufferin Street, Downsview, Ontario, M3H 5T4, Canada. Other Canadian weather information is also available from this source.

Be aware that weather conditions at altitudes of only a few thousand feet can differ drastically from those at local weather stations where climate data are collected; temperatures generally drop 3° to 4°F for every 1,000 feet of elevation, and winds—and therefore wind chill—can increase dramatically. Always take altitude into account when estimating weather conditions.

PRETRIP PLANNING

Choosing Equipment

Having thoroughly investigated the weather conditions in your field area, you must now choose the clothing and equipment you'll need to work in that environment. Being properly equipped is probably the single most important step you can take to increase your chances of safety in the field.

Essential Gear

The following is a recommended list of basic equipment that you will need. The list should be expanded or shortened to meet the needs of your particular trip.

- ☐ Waterproof ground cloth; lightweight tarp for emergency shelter, if fluorescent or bright color, can also serve as signal cloth for aircraft; or heat-reflecting "space blanket"
- ☐ Reliable, waterproof flashlight with extra batteries
- ☐ Filled water bottle or canteen
- ☐ Water purification tablets or water filtration device
- ☐ Lightweight nylon rope, about 30 feet long
- ☐ Waterproof matches or strike-anywhere matches in a durable waterproof container
- ☐ Fire-starter jelly, butane lighter, magnesium block, slow-burning candle, or equivalent campfire aid
- ☐ Pocketknife, pliers, or multifunction camping tool

PLANNING FOR FIELD SAFETY

☐ Compass
☐ Large-scale topographic map or air photos of the area
☐ Extra food, at least a day's worth; high-calorie, ready-to-eat
☐ Extra warm clothing and/or rain gear
☐ Distress whistle (much more effective than shouting for help) or signal-flare pen
☐ Signal mirror (often part of compass)
☐ Sunscreen and sunglasses
☐ Medical kit

Clothing

From the standpoint of safety, the primary function of clothing in the field is to keep the wearer warm, dry, and protected from the sun. For information on types and amounts to bring, see "Cold Weather." Fieldworkers heading to normally warm locales should also consult that section; even the hottest climates can sometimes present dangerously wet and chilly conditions, particularly at night or at high altitudes. In general, try to select brightly colored clothing to promote visibility in case you get lost.

Footgear

All field party members should determine the proper footgear needed for the activities that are planned. Avoid

PRETRIP PLANNING

the temptation to use those cracked leather boots with the worn-down soles for "just one more season."

If you buy new boots for your trip, make sure they fit properly—too tight and toes will become cramped and easily chilled, too loose and you'll soon have painful blisters. When sizing boots, be sure to wear the same number and type of socks you'll be wearing in the field. Break in your boots by taking short hikes around the neighborhood; don't wait to break them in on the trail. If your boots are leather, waterproof them before heading out. If you'll be out for several weeks, it's a good idea to bring waterproofing compound with you so you can reapply it in the field. If faced with stream crossings, you may want to carry an old pair of jogging shoes or plastic sandals to keep your hiking boots dry.

Work Gloves

Bring a couple of pairs of heavy leather or canvas work gloves to protect your hands from cuts, abrasions, and blisters during fieldwork.

Tools

Choose your field tools carefully, based on the type and amount of work you'll be doing. Trying to make do with a tool that was not designed for a given task is inefficient and can be dangerous.

Examine all tools before you head out to make sure they are in proper working order. If you're not familiar

PLANNING FOR FIELD SAFETY

with the operation of a particular tool, get instruction from a qualified person before you attempt to use it. Practice using the tool to develop proper, safe technique. See "Equipment Precautions" for more specific safety instructions.

Rope

Because of its strength and light weight, parachute cord is a good choice for securing food in trees, tethering boats, and tying down items.

Backpacks

If you'll be carrying a backpack, make sure the frame (external or internal) is the correct size for your body and that all straps and harnesses have been adjusted properly. An internal frame is better in wooded areas to avoid getting hung up on limbs. An external frame may be preferable in open areas for comfort and to help distribute the weight of a heavy load evenly. Also make sure that your pack has a quick-release belt buckle so you can take the pack off quickly in an emergency.

Don't overload your pack. An extremely heavy pack makes you more likely to lose your balance during river crossings and on treacherous terrain.

Take along materials to repair your pack.

Tents

With the exception of those few areas of the globe where the weather is warm and dry and insects are few, you'll probably want to use a tent if you're not staying in a cabin or hotel. As with most equipment, the type of tent you bring will depend largely on the weather conditions you expect to face (i.e., don't bring a two-season tent—generally taken to mean spring and summer—if the two seasons you plan to be out are fall and winter) and whether you'll be packing the tent to your base camp (in which case a light, nylon tent is best) or transporting it there by vehicle.

Because wind is often the chief enemy of a tent, choosing an aerodynamically stable model is advisable. Loss of a tent in a storm can create a life-threatening situation. Brightly colored tents are easier to locate in the fog or clouds by returning fieldworkers or by rescue aircraft.

Practice setting up your tent before heading out to make sure you have all necessary equipment, such as poles and stakes and anchors for sand or snow. Standing in the dark and rain amidst a jumble of tent hardware is no time to discover that poles are missing and assembly instructions are written in a foreign language. If you're heading into a remote area, consider bringing some extra tent pegs and poles to replace those that get lost, bent, or broken. .

A sponge and a small whisk broom will help keep the interior of your tent clean. Keeping a tent free of food debris or spills will help deter damage by vermin or

PLANNING FOR FIELD SAFETY

bears. The seams of a tent should be sealed every couple of years to keep it waterproof. It is advisable to waterproof your tent before heading into a rainy field area. In an area where insects are abundant, you may want to bring insect netting.

Stoves

Make sure your stove is in good working order and that you know how to operate it safely before heading into the field. For information on the different types of stoves available and the safety precautions that apply to each, see "Stoves" in chapter 2.

Campfire Starters

Although campfires are now sharply restricted in many areas, being able to light one quickly in an emergency, whether to use for heat or to signal for help, can mean the difference between life and death. Brush up on your campfire skills before heading out (most camping books cover the basics), even if you'll be working in an area where campfires are forbidden. Be sure to bring along a tube of fire-starter jelly, a magnesium block, a slow-burning candle, a butane lighter, or some other fire aid to speed the process of starting a campfire, particularly when using wet materials.

For campfire safety guidelines, see "Campfires."

PRETRIP PLANNING

Medical Preparations

It is advisable to consult a physician when determining your medical needs.

Immunizations and Medications

Before starting your trip, investigate the area where you will be working so that you can answer the following questions: Are immunizations required? If so, which ones? When do you need to begin the immunizations? For information on immunizations for foreign and domestic travel, see a current copy of *Health Information for International Travel,* U.S. Public Health Service, Superintendent of Documents, U.S. Government Printing Office, Washington, D.C. 20402. Immunization information is also available from the Centers for Disease Control at 404/639-3534.

Are there any medications you'll be required to take while in the field (e.g., antimalarials)? If so, make sure you have an adequate supply.

Medical History

A medical history form should be filled out by all party members and a confidential copy kept with the base camp medical kit. Important information includes current physical condition; current state of health; allergies to drugs, insects, foods, or other substances; dietary

PLANNING FOR FIELD SAFETY

restrictions (e.g., lactose intolerance); high or low blood sugar; reactions to temperature extremes; reactions to altitude; high or low blood pressure; heart conditions; physical handicaps; and diabetes.

Personal and Base Camp Medical Kits

Plan your personal medical kit and base camp medical kit based on the local terrain, altitude, season, climate, the length of the trip, the size of your group, the medical training of the group leaders, and any problems specific to the area/ecosystem (e.g., desert: dehydration, heat exhaustion, heat stroke, snakes, scorpions; high altitude: hypothermia, altitude sickness; rain forest: chronic athlete's foot). Consider the medical problems that are likely to occur, and carry the equipment needed to provide effective first aid treatment until the victim can, if necessary, be evacuated safely to a medical facility.

Personal medical kit. Each person in the party should carry a personal medical kit if he or she is apt to be separated from other party members. (If a small group will be in the field together, a small group kit could be carried instead of individual kits.) These kits should be tailored to the locale as described above. They should be expanded if the individual or group will be away from camp for more than one day. Some items that might be found in such kits include the following:

PRETRIP PLANNING

- ☐ First aid book
- ☐ Pencil and paper
- ☐ Adhesive bandages and bandage strips, assorted sizes
- ☐ Adhesive foam for blisters (moleskin or Spenco adhesive knit)
- ☐ Roll of 2" wide adhesive tape
- ☐ Triangular bandages (also known as cravats)
- ☐ Several 3" x 3" sterile gauze pads
- ☐ Roll of gauze
- ☐ 3" wide elastic bandage (Ace bandage)
- ☐ Alcohol swabs
- ☐ Neosporin or other triple antibiotic ointment
- ☐ Buffered aspirin or other pain reliever
- ☐ Water purification tablets or kit
- ☐ Sterile eyewash
- ☐ Sterile eye patches
- ☐ Sunblock
- ☐ Lip balm
- ☐ Safety pins
- ☐ Steel sewing needle
- ☐ Razor blade
- ☐ Small bar of antibacterial soap for cleansing wounds
- ☐ Small scissors

PLANNING FOR FIELD SAFETY

- ☐ Tweezers with needle point (may be part of pocketknife)
- ☐ Personal medications
- ☐ Patient Assessment Form

Base camp medical kit. The base camp medical kit is an expanded version of the personal medical kit and should be modified according to expected field conditions.

The base camp medical kit should contain an up-to-date first aid book, providing detailed information on medical procedures in the field. Two useful books are *Medicine for Mountaineering*, edited by James A. Wilkerson, and *Medicine for the Outdoors: A Guide to Emergency Medical Procedures and First Aid*, by Paul S. Auerbach. All field personnel should become familiar with the book's contents and organization so that required information can be found quickly in an emergency. Field personnel should also be certified in CPR and first aid.

The amount of each item in the base camp medical kit will depend on the size of your group and the length of your trip. More sophisticated supplies can and should be added if group leaders are trained in their use. Starred items are optional. All items should be kept clean and dry.

- ☐ All items in the Personal Medical Kit (increase amounts as necessary)
- ☐ Sterile absorbent cotton

PRETRIP PLANNING

- ☐ Sanitary napkins
- ☐ Nonstick sterile gauze pads
- ☐ Butterfly bandages
- ☐ Thermometer
- ☐ Iodine prep pads
- ☐ Povidone iodine, Betadine, or other antiseptic
- ☐ Petroleum jelly
- ☐ Glucose
- ☐ Granulated salt
- ☐ Rehydration mix (glucose + salt water)
- ☐ Antacid tablets
- ☐ Laxatives
- ☐ Antifungal cream
- ☐ Nasal spray
- ☐ Throat lozenges
- ☐ Cough medicine
- ☐ Antidiarrhetic
- ☐ *Instant chemical cold pack
- ☐ *Instant light stick
- ☐ *Nail polish (for chiggers)
- ☐ *Wire-mesh splint, SAM splint, or inflatable splint
- ☐ *Rescue blanket

PLANNING FOR FIELD SAFETY

☐ *Roll of duct tape

If you are bringing prescription medications, it's advisable to include a back-up supply in the base camp medical kit. Field personnel should know when and how to use the contents of the medical kit. An emergency is not the time to learn skills and to use unfamiliar equipment.

Supplementary Medications

If you are going into a remote area, you might consider taking the following medications. Discuss their use with a physician. Some are prescription drugs.

☐ Topical anesthetic
☐ Topical anti-inflammatory medication
☐ Eye antibiotic
☐ Eye anesthetic
☐ Strong analgesic
☐ Broad-spectrum antibiotic
☐ Decongestant
☐ Medications for allergic reactions, motion sickness, nausea, vaginitis, and acute mountain sickness

Medical Kit Guidelines

Label all medications clearly. Include dosage information and directions for use.

Check the expiration dates of all medications before you leave, and replace those that have expired or will expire during your trip.

Keep all supplies together in a clean, dry, cool place. Seal all instruments, gauze, cotton, and other items that must remain sterile in plastic bags. Heat-sealed food bags work well.

Do a periodic check of the kit to make sure essential supplies are in order.

Replace depleted supplies as soon as possible.

Keep a logbook showing what important medications were used, and when.

Provide each van on a field trip with its own medical kit in addition to base camp and personal medical kits.

If you're working in wet areas or on water, be sure medical supplies are kept in a waterproof container.

Food

Fieldwork is strenuous, and fieldworkers need plenty of nutritious food. In cold weather, a fieldworker needs about double the amount of food that he or she needs in hot weather. A cold-weather diet should also be higher in fats and oils. It's a good idea to bring enough food for at least one extra day.

PLANNING FOR FIELD SAFETY

Plan menus for each meal before heading out. Lightweight, nutritious foods include pasta; powdered mashed potatoes, milk, sauces, and soups; biscuit mix; oatmeal; gorp (good old raisins and peanuts); instant desserts; and rice. For easy transportation and preparation, many people now opt for the more expensive freeze-dried foods, which are also less likely to attract animals because the foods have little odor.

Check the medical history of each group member before heading out to make sure there will be sufficient acceptable food for those with food allergies, lactose intolerance, or other dietary restrictions.

Logistics

Before arriving in an unfamiliar area, learn as much as you can about the environment, culture, regulations, and expenses. If you can't find an individual who has previously done fieldwork in your area, interview someone who has been a tourist there. Do your research well before your departure. Personal communication is your most detailed source of information.

Always set up a check-in system with a reliable person before you leave on any outing, whether you're going on an extended trip into the backcountry or simply a day trip away from base camp. Tell the person your planned route, possible alternate routes, and expected time of return. If you have a radio, set up a daily contact schedule (usually morning or evening), and be sure to carry extra batteries. Establish a margin of error for your return time

PRETRIP PLANNING

or radio contacts, after which a search party should be sent out to find you (using the itinerary information you provided).

It's also a good idea to leave a copy of daily travel plans, including departure time and pertinent maps, in a conspicuous place in camp, to further aid searchers.

You should have a general search procedure planned before a problem arises. An excellent resource for search and rescue is *Wilderness Search and Rescue* by Tim Setnicka.

Plan traverses with the sun at your back, and allow ample time so that you can get back to camp while it's still daylight.

Plan your trip so that light work days are interspersed with hard ones. Avoid scheduling consecutive days that require maximum physical output.

If you have student assistants, make sure that they have carefully explained their responsibilities and the trip conditions to their parents.

If you'll need to work on private land, obtain permission well in advance. Explain to the landowner the nature and purpose of your work, and indicate what heavy equipment you'll be using, if any. Make sure the owner is fully aware of any digging, blasting, sample removal, or other work that will disturb property, and arrange suitable compensation before work begins.

Except for restricted areas, such as military bases and nuclear test sites, you generally don't need permission to work on federal land. You will, however,

PLANNING FOR FIELD SAFETY

need permission to dig, remove samples, or otherwise disturb lands in national parks and monuments.

Rules for working on state-owned land vary from state to state. Investigate the rules in your state before you begin work.

Evacuation Plan

Formulate an evacuation plan to cover the possibility of a serious injury. The plan should include general procedures for evacuating the victim to a medical facility and information on other rescue resources in the area. Contact local rescue workers for help in tailoring the plan to the specific region and to coordinate potential rescue efforts. See "Evacuation Procedures."

Personal Preparation

The field is not the place to get into shape. Work on developing good aerobic and muscular conditioning before you head into the field. You will become more efficient in the field, and your increased strength, endurance, and agility will reduce your likelihood of accidents, most of which occur when people are fatigued and using poor judgment. Being in shape will also make your trip more enjoyable.

If there's even the slightest chance you'll be in water—crossing a stream, flying over a lake, camping on a beach, taking a bath in a river—know how to swim.

PRETRIP PLANNING

Know modern first aid and CPR procedures thoroughly; lack of this knowledge is a serious disability. Contact your local American Red Cross chapter or other health organizations for information on training courses.

Know when and how to use the contents of the medical kit. Read through (and periodically review) your first aid book to learn what information it contains and how it's organized. You must be able to find pertinent information quickly in an emergency.

If you have any unusual medical problems, consult a physician before your trip, and bring any necessary medications or equipment. If you'll be taking prescription drugs, consider bringing a duplicate supply for safe storage. It is advisable that you inform some members of the field party of your medical problems and where your medication is stored should you become incapacitated.

If you'll be out in the field for several weeks, it's a good idea to have a dental checkup before you leave; a cracked or infected tooth can cut short your trip.

If you wear eyeglasses or contact lenses, carry an extra pair, along with your eyeglass prescription. Also bring extra sunglasses. Unless your glasses are the type that hooks around the ears, wear a restraining strap to keep them from falling off and breaking.

Find out what emergency medical resources, such as the U.S. Forest Service or a local hospital, are available in your area and how to reach them quickly.

Knowing how to build and light a fire can save your life in an emergency; make sure you know the basics.

PLANNING FOR FIELD SAFETY

Most camping books can help you brush up on this essential skill. *How to Stay Alive in the Woods* by Bradford Angier is a classic reference that covers this and other wilderness survival skills. Consider carrying the book or a similar one with you on your trip.

Safety Regulations

Employers and employees should be aware of the field safety regulations of the Occupational Safety and Health Administration (OSHA) and the Environmental Protection Agency (EPA). Section 29CFR1910.120 of the OSHA regulations specifies that people who are engaged in field activities where hazardous waste may be found (e.g., a Superfund site) are required to have 40 hours of health and safety training, and their supervisor must have an additional eight hours of training. Employee training must be equivalent to the Red Cross Standard First Aid course, or employers are subject to fines or other penalties. Section 40CFR264.16 of the EPA regulations specifies that personnel at hazardous-waste facilities must successfully complete a program of classroom instruction or on-the-job training that teaches them to perform their duties in a manner that ensures their safety.

2

EQUIPMENT PRECAUTIONS

Tools and Equipment

Tools

Before using any tool, make sure it's in safe working condition—head firmly attached, handle intact, blade sharp. Wear heavy work gloves to protect your hands from cuts and blisters. With some tools, such as axes and chainsaws, it's wise to wear boots with steel-reinforced toes. If you're unsure of how to use a tool safely, have a qualified person instruct you.

Rock hammers. Always wear eye protection when using a rock hammer. Never strike two rock hammers—or any other hardened steel tools—against each other; they could shatter. If the hammer head becomes mushroomed, file down any sharp burrs. Never use the rock hammer's pointed end as a hammer; it is designed for prying only.

A chisel is useful for concentrating the force of a rock hammer and for carefully excavating fossils, well-formed mineral crystals, and other delicate samples. If you think you'll need such a tool, bring one. Never use your rock hammer's pointed end as a chisel, striking the flat end with a second hammer; such actions are extremely dangerous.

Cutting/chopping tools. Keep tools such as knives, saws, axes, and machetes sharp. Dull tools require more cutting force and tend to slip on wet wood and glance off knots, greatly increasing your chances of injury. Be sure to bring a small sharpening stone with you, and use it as necessary.

Hatchets are much more likely to cause accidental injuries than axes are. Their one-handed operation is unstable, they require extra force to cut effectively, and their cutting head is very near the user's body. Hatchets are also less efficient than axes, and because they often differ from axes in shaft size only, they are not significantly lighter. If you think you'll need a hatchet, bring an ax.

When using any chopping tool, always wear eye protection and sturdy boots; keep your body clear of the blade's follow-through; and make sure you're well clear of other members of your group. Always hold an ax firmly with both hands, and make sure your feet are firmly planted before you start swinging.

When using a machete, always use the wrist loop to keep the tool from flying out of your hand.

Keep all sharp tools in their protective sheaths when not in use.

Chainsaws. Chainsaws can be extremely dangerous tools if improperly handled; use only if you are thoroughly familiar with their safe operation.

EQUIPMENT PRECAUTIONS

The cutting chain must be kept sharp, oiled, and tight (but not too tight—follow the manufacturer's recommendations) for safe, efficient cutting. Bring a sharpening file and use it as necessary. Periodically tighten the chain throughout the day, since heat generated by friction with the wood will tend to expand and loosen the chain. If your chainsaw does not oil the chain automatically, remember to do so manually by depressing the oil button periodically as you cut.

Watch for dangerous kickbacks, which occur when the chain on top of the cutting bar comes in contact with the wood or other object; cut only with the underside of the cutting bar.

Never attempt to cut timber larger than that for which your chainsaw is designed; match your tool to the task.

Use caution when fueling your chainsaw. See "Stoves" in this chapter for safe handling of fuel.

Always wear safety goggles; chainsaws spit out woodchips—and anything else they contact—at great speed. Kevlar safety chaps are available for leg protection. To prevent hearing damage, wear ear plugs or other ear protection. A hard hat may be required when using a chainsaw in some areas.

Compass. Protect a compass from the elements and from excessive jarring. Be careful not to hold the compass near local sources of magnetism—such as metal fence posts, belt buckles, clipboards, under power lines—when taking readings. If your map and compass skills are a bit rusty, brush up before heading out. Practice methods of triangulation so you'll be able to pin-

PLANNING FOR FIELD SAFETY

point your location if you get lost. Remember to adjust your readings for the local declination. *Be Expert with Map and Compass,* by Bjorn Kjellstrom, is an excellent reference.

Maps. Unless you're working in a desert environment, have maps waterproofed or keep them in a waterproof pouch. Zip-lock bags work well. Make sure that each separate field team is carrying a local, detailed topographic map.

Stoves

Whichever type of portable cooking stove you use—white gas, kerosene, butane, or alcohol—remember that each can pose serious dangers if improperly operated. Learn to use your stove before heading into the field, test it to make sure it's in good working condition, and use the same care around your stove that you would around any open flame. Keep your campsite free of flammable debris. Set up your stove on a stable, nonflammable surface, such as a flat rock, small stones, or bare soil. Do not place the stove directly on grass or other vegetation, which could dry out and ignite. And never—no matter how miserable the weather outside—use any stove inside a tent. Even if you manage to avoid burning your tent to the ground, the buildup of carbon monoxide inside the tent can kill you.

White-gas stoves. The most commonly used type of stove in the United States, white-gas stoves are also potentially the most dangerous. A highly volatile fuel,

EQUIPMENT PRECAUTIONS

white gas will ignite at the slightest spark—and the stoves keep this fuel under very high pressure.

Never overfill the fuel tank: fill only to the level recommended by the manufacturer. Proper filling is especially important in cold weather when the heat of the stove will cause previously cold fuel to expand greatly.

After each filling, completely tighten the stove's fuel cap. Wipe up all fuel spills, and move the stove to a new location before lighting.

Keep your face away from the stove when lighting, in case flames flare up.

Never set up a windscreen that completely blocks ventilation around the stove; with a large pot or pan on top of the stove, such a screen can cause reflected heat to build up to explosive levels. Although the stoves are equipped with a pressure-release valve to prevent such explosions, serious blowouts have been known to happen.

Never use automotive fuel in place of white gas; additives in the fuel will in time clog the stove's fuel lines and increase the chance of a blowout.

Take care not to spill white gas onto your bare skin in subfreezing temperatures; because the fuel evaporates quickly, it can cause instant frostbite.

When carrying fuel, use metal or plastic bottles designed for that purpose, not plastic water bottles, glass jars, or other containers. Never leave full fuel bottles in the sun; they can expand and rupture. To be on the safe side, leave at least one and one-half inches of space in each fuel bottle to allow for expansion.

PLANNING FOR FIELD SAFETY

Kerosene stoves. Use a kerosene stove (or a multifuel stove that can burn kerosene) if you'll be doing extensive fieldwork in a foreign country. White gas is often difficult to find outside the United States and Canada, particularly in the more remote regions of the world.

Kerosene stoves are also more dependable in extremely cold conditions and at high altitudes. Although the fuel doesn't burn as cleanly as white gas, it is considerably less volatile than white gas, and thus will not readily ignite when spilled.

Butane stoves. Butane stoves operate on self-contained, disposable butane cartridges and, while not as powerful or efficient as white-gas and kerosene stoves, are generally safer and easier to use. But don't bring a butane stove if you'll be working in an area where temperatures can approach freezing: as temperatures drop, the butane cartridges lose pressure and the stove's heat output falls; below about 35°F, butane stoves won't work at all.

Alcohol stoves. These stoves burn denatured alcohol. Advantages of ethanol fuel include easy lighting (no need to prime the stove), good operation at below-freezing temperatures, and low volatility. The flame is not as hot nor as clean as a white-gas flame.

Gas Lanterns

Burn new lantern mantles for at least 15 minutes in open air before using the lanterns in an enclosed space.

EQUIPMENT PRECAUTIONS

New mantles release thorium and beryllium gases, which are poisonous.

Hang lanterns securely and far enough away from the walls and roof of your tent to eliminate any risk of fire. Take care when moving a lit lantern, as handles can get dangerously hot.

Firearms

Firearms should only be brought into the field to protect fieldworkers. Be sure the size and caliber of a weapon matches the risk. Require proof of proper firearm training before allowing anyone to bring or carry firearms. Obtain training yourself. Don't allow anyone not familiar with their safe operation to touch firearms. Lay down ground rules concerning their safe handling and use before a mishap occurs. Mishandling by either the person in charge of the weapon or the inquisitive can lead to fatal accidents.

Store firearms and ammunition together so they can be found quickly, but do not load firearms until you need to use them.

In areas where you need loaded firearms for protection, exercise extreme caution at all times. Load the magazine, but keep the chamber empty. Don't forget to chamber a shell before attempting to fire the weapon.

If it is necessary to kill a dangerous animal in self-defense, shoot to kill, and document the complete circumstances leading to the kill: actions of the animal; attempts to scare it away; distance of the shot(s); name of the shooter; and photographic documentation. Report the incident immediately to authorities.

Treat all firearms as though they are loaded.

Never point a firearm at anything you do not want to kill, and be sure of your target before pulling the trigger.

Never use firearms if you are under the influence of alcohol, medications, or other drugs.

Keep all firearms clean and in proper working order.

If you're planning to transport firearms on commercial airlines, check with your airline beforehand for specific restrictions.

Drilling Rigs and Sites

Drilling rigs range from the large stationary rigs used for deep drilling to recover oil and gas, to small portable one-person units used to obtain shallow samples. In either case, a drill rig is inherently dangerous because it has a rotating drill stem or auger that can quickly break a bone or remove part of a body. The International Drilling Federation publishes an excellent booklet titled *Drilling Safety Guide*. This should be standard reading for anyone who will be working around drill rigs.

Safety around a drilling operation is paramount to anything else. Never enter the working platform or approach a truck-mounted drill rig closer than the height of the mast without the knowledge and permission of the drill rig operator. Never smoke near an oil and gas exploration rig or near a rig working around environ-

EQUIPMENT PRECAUTIONS

mental sites, such as landfills or underground storage tanks, where explosive gases may be encountered. If smoking is allowed, be sure that your cigarette is out when you dispose of it to prevent a rig or grass fire. Never put your hands on the controls or on the rig. Keep away from all moving parts or parts that may move at any time. Never touch the cables on the rig; they may move without notice.

Geologists are often more interested in the samples recovered than in safety, so don't let your excitement about a new find break your arm. A safety supervisor should be appointed for each drill rig. On stationary rigs and truck-mounted rigs this person is usually the driller or the drilling superintendent. All workers and visitors around the rig are required to follow the instructions of the safety supervisor and are responsible for their own actions.

The driller has the full responsibility for the safety of the drilling operation, the drill rig crew, visitors, and for the safe condition of the drill rig. The driller or the safety supervisor has the authority to enforce all safety rules and regulations. The driller can, and will if necessary, have a visitor or crew member removed from the site for failure to follow safety procedures.

In some cases, the geologist will be responsible for the samplers as they are recovered. Always follow the instructions of the driller, and be sure that you have been trained on the rig on which you are working. A few minutes of training by the driller on the activities of the drilling cycle, the characteristics of the rig, and the

PLANNING FOR FIELD SAFETY

drilling procedures will help maintain safe operations at the site.

All drilling operations must comply with environmental and safety regulations, including those of the Mine Safety Administration (MSA), the Office of Surface Mining (OSM), and the Occupational Safety and Health Administration (OSHA), and local regulations. In some states the driller must be licensed. Never ask a driller to perform a drilling service that is not included on the license. Regulatory boards have held the geologist or engineer who hired the driller liable for a fine for "operating a drill rig without a license" because the driller was not informed of the purpose of the hole or well and was not licensed to perform the service required. Be sure to review the regulations that are in effect in the area where you plan to have drilling services performed or where you plan to do your own drilling. You could be fined for "drilling without a license."

Individual Safety Equipment

Safety head gear. Safety hats (hard hats) must be worn at all times by anyone working or visiting a drilling operation. Safety hats must comply with the requirements of the American National Standards Institute (ANSI) Z89.1.

Safety shoes or boots. Safety shoes or boots must be worn at all times by all drilling personnel whenever they are within "close proximity to a drill rig." Close proximity is a distance equal to the height of the mast on the rig.

EQUIPMENT PRECAUTIONS

Safety shoes or boots should provide at least toe protection. Around heavy equipment, the instep should also be protected. Wear the right boots for the ground conditions; smooth soles can become very slippery in wet or muddy conditions. Never wear leather-soled shoes or boots. All safety shoes or boots must meet the requirements of ANSI Z41.1.

Safety glasses. Drilling personnel and visitors that come within close proximity to a drill rig must wear safety glasses. The safety glasses must meet the requirements of ANSI Z87.1.

Safety gloves. Wear gloves when working around a drill rig. Most drillers use a cotton glove with "buttons" to assist in maintaining a sure grip on wet, greasy, or muddy items. Replace gloves whenever they become too slippery to safely handle tools and samplers. Most safety supervisors will not allow leather gloves around the drill site because leather gloves can become very slippery when they are wet. Never wear gloves that are loose or have a cuff because they may get caught in the rotating parts of the rig.

Hearing protection. Noise levels around some drill rigs can be excessive, especially around some of the small portable units. Always wear hearing protection in high noise level areas. The best type of protection should be designed to fit in your ear. Bulky ear muffs and head gear can get in your way and cause accidents.

General clothing. If you will be working as part of the drill crew, always wear clothing that fits tightly on your body. Loose clothing can easily be caught in the rig and cause significant injury or death. Never wear shorts. Don't wear jewelry, especially neck chains, earrings, or rings; they can be caught or crushed. Never roll up the sleeves of a shirt and always keep all cuffs buttoned.

Other protective equipment. Special protective equipment must be used and worn in certain drill site conditions, such as hazardous materials sites. The equipment and instruction in its use may be provided by the project manager or the safety supervisor. If they are not provided, you are responsible for meeting the safety requirements. Some drilling conditions also require that the drill crew and visitors be specially trained to use special protective equipment before entering a site. Never enter a site that requires special training or equipment unless you have had the training and are current. Entering such a site could result in a significant fine as well as a safety risk to all personnel on the site.

Hand Tools

In many cases, the geologist will be responsible for handling the sampling devices, such as core barrels, thin tube samplers, and split spoon samplers. If you are assigned responsibility for the samplers, be sure you know how they work. If not, ask the driller to demonstrate safe handling procedures before you work with the samplers. You will be required to use simple hand tools, such as screwdrivers, hammers, and pipe wrenches, to open the sampler or hydraulic extruders to remove the sample.

EQUIPMENT PRECAUTIONS

Use each tool only for its intended purpose. "Use the right tool for the job" is the best safety rule that you can follow. When using a hammer you should wear safety glasses. Be sure that the teeth on a pipe wrench are clean of grease and mud before using the wrench. Clean the teeth with a wire brush.

Watch where you place your fingers. Do not allow them to be pinched between tools or between a tool and the ground or rig. Be sure that your hands are clear of any hydraulic sample extruder when placing the sample tube in it.

Drilling Site

The geologist often selects the preliminary location of the drilling site to meet the objectives of the exploratory program. To make sure the site is safe, review it with the driller. Be ready to relocate if conditions in the field are unsafe or if the driller determines that the site is unsafe. Never ask a driller to set up where trees will interfere with the mast or where the ground is unstable or on steep slopes. Always check for overhead wires. Never assume that the line is dead! A residential power line carries 440 volts and is capable of burning or killing if it arcs to the mast. Always check with local and regional utility companies and cities for the presence of underground utilities on a proposed drill site.

Whenever a drill site is located near overhead power lines, the mast must be at least 100 feet from the wires. High voltage transmission lines may require greater distances to ensure against arcing to the mast. Lesser distances from overhead lines can be reviewed with local authorities; however, all distances must comply with OSHA Reg. 29CFR1910.180.

Never allow the rig to be moved with the mast raised. It can contact overhead wires or cause the rig to topple. Stay clear and in sight of the driver whenever the rig is being set up over the drill site. If you are directing the driver while setting up over a drill site, always stand so that you can see the driver's face in the truck mirror.

Stay alert for the movement of support equipment, such as the water truck and supply/service truck when working around a site. Always park your vehicle in a safe place. Park at least a distance of more than the height of the mast from the rig.

When making field descriptions of the cores/samples and when sampling the core, always set up your workplace more than one mast height from the rig, unless a geologic laboratory has been provided at the site. Visitors interested in the geology are therefore out of close proximity to the rig and do not need special protective equipment. Do not allow visitors to approach within close proximity to the rig. You are responsible for their safety.

If you are doing your own drilling, always know how to safely operate the equipment. You are the safety supervisor and are responsible for all aspects of the operation. Be sure that your helpers are trained and that they know their jobs. Never attempt to use a drill rig for a purpose that it was not designed to handle. Forcing a

EQUIPMENT PRECAUTIONS

rig beyond its design capabilities can and will cause accidents!

3

SAFETY IN THE FIELD

Fieldwork

When you strike out into the backcountry, you trade the familiar hazards of civilization for the relatively unknown hazards of the field. Although many of these hazards can be life threatening, nearly all can be avoided or safely handled by exercising sound judgment at all times, taking necessary precautions, remaining alert to potential dangers, and acting quickly and effectively to counter these dangers before they become serious problems. This chapter contains precautions for working and camping in the field safely.

Consider all aspects of the environment, such as wind, visibility, precipitation, temperature, cloudiness, barometric trends, and tides, when making field plans and decisions. For example, hot weather can turn a snow-fed stream that was easy to cross in the early morning into an impassable torrent by late afternoon, and rising tides can cover secure-looking sandbars.

Plot your routes carefully. Be realistic about how far you can travel in the time available. Plan an alternate, shorter route to follow if you find yourself unable to complete the original route. Careful planning is essential

PLANNING FOR FIELD SAFETY

to prevent situations where separated field groups are forced to do something dangerous to rejoin the main party.

Try to schedule difficult activities when support is close at hand. For example, if you find yourself separated from your party and need to climb a rock wall or cross a deep stream, don't do it until other people are around.

Plan to get back to camp earlier rather than later than necessary, that is, before dark or to meet a radio contact schedule. Accidents are much more likely to occur when you're moving in haste or tired.

Whenever you travel any significant distance away from the base camp, carry the items on the essential gear list.

Think twice before taking a dog into the field. Dogs sometimes can't keep up with humans over rugged terrain—basalt flows, snow fields, talus slopes—because of tender, unprotected feet; they can incite attacks from wild animals and provoke unfortunate encounters with porcupines and skunks; they can run off, causing you to waste time looking for them; they can knock rocks onto your head from overhanging outcrops, and otherwise hamper your fieldwork; and they require you to carry extra food and water. Remember that any dog taken into the field must be inoculated for rabies. In most cases, dogs should not be taken to the field. See also "Bears."

A useful field guide for those working in rugged terrain is *Mountaineering: The Freedom of the Hills*, edited by Ed Peters. This is considered one of the best, most comprehensive books on the subject and is especially good for those new to fieldwork.

Marijuana and Moonshine

Beware of marijuana plots and farms anywhere in the world where the climate is warm enough for the plant to grow. The owners will often defend their crop violently, sometimes shooting first and dispensing with questions altogether. Avoid the plots. If you stumble onto one, don't linger. In the southeastern United States, moonshiners may still defend their territory just as violently, so treat them with equal caution.

Emergency Plans

Discuss emergency procedures with the other members of your group before setting out. Always plan for the possibility that you may not be picked up or be able to return to camp because of weather, mechanical breakdowns, or accidents. If distance, a sizable stream, or other obstacles could prevent you from returning to camp on foot, be sure to carry enough clothes, equipment, and food to make an emergency bivouac overnight without risk of hypothermia. Periodically check emergency equipment to make sure that it's functional and in order (i.e., are emergency matches wet? Is bandage supply running low?).

An alternative to carrying emergency equipment is to drop off an emergency supply cache in a location that

PLANNING FOR FIELD SAFETY

you will be able to get to easily. Mark its location on your map and flag it with flagging tape or a signal cloth. Make sure the cache includes enough equipment for the number of people in your group. If you're working in an area of permafrost, be sure to place the cache on a rock, not on the ground. Otherwise, the weight and warmth of the cache will melt the permafrost, causing the cache to sink into the ground and freeze.

If you plan to use an emergency cache, it should be carefully packed and checked before you head into the field. Be sure to locate the cache out of sight of heavily traveled trails to minimize the risk of theft.

An emergency cache should include the following items:

- ☐ All the gear listed under "Essential Gear." In cold weather, carry a small tent in place of the emergency tarp.

- ☐ Sleeping bag. Use one filled with synthetic insulation, which will provide warmth even after a soaking.

- ☐ Fluorescent signal panel at least 4 feet x 4 feet and perhaps some small smoke flares. Signal devices become more necessary if the person picking you up is not the one who dropped you off.

- ☐ Ensolite pad and an extra set of polypropylene or wool long underwear if working in high mountains, ice fields, glaciers, or anywhere the ground is apt to be cold.

SAFETY IN THE FIELD

- [ ] Ax, small crosscut saw, machete, or equivalent cutting tool if working in timbered or brushy country.
- [ ] Ice ax, climbing rope, and other climbing equipment, if conditions require them.
- [ ] Water or a means to melt snow in areas where water is unavailable or frozen.

Transportation

Every van on a field trip should carry a basic medical kit.

Always lock vehicles when unattended, and make sure all valuables are out of sight; robberies are not uncommon in the backcountry.

When ferrying fieldworkers to a site, make sure a supply of food, shelter, medical equipment, and other emergency gear goes with the first group, just in case bad weather or breakdowns prevent the rest of the party from joining them that day.

If a vehicle is going to meet you, make sure the driver or pilot knows the rendezvous point. Also set up an alternate meeting place in case you can't get to the first one.

If you are set down by helicopter onto a gravel bar, make sure the pilot is instructed to wait until your entire party is safely ashore before flying off.

PLANNING FOR FIELD SAFETY

Keeping Track

A good procedure, regardless of your field area, is to keep a daily logbook indicating your planned route for each day, possible alternate routes, and your estimated return time. Leave the logbook in camp in a sheltered place where searchers would be likely to find it if you were to have an accident. Such a system should be a back-up to the regular check-in system discussed in "Logistics."

Use your map to keep track of your position at all times. Don't allow yourself to go miles out of your way before realizing your error. If you decide to follow an unmapped logging road or a game trail, check your heading constantly to be sure you're not being steered off course.

Protective Gear

Wear eye protection when hiking through dense brush.

If you're working in the woods during hunting season (not a recommended practice), be sure to wear a fluorescent vest and other bright garments. Some states, for example, Maine and Massachusetts, require this.

Wear a hard hat when working in caves, mines, road cuts, or other environments where falling debris is a possibility.

SAFETY IN THE FIELD

Risky Conditions

If you ever find yourself unprepared for the climate, terrain, or other field conditions, turn back immediately. No fieldwork is worth risking your life.

Don't do anything that looks dangerous to you. Your chances of an accident will be greatly reduced if you stay within the range of your physical abilities and your climbing experience. Group leaders should remember to limit activities to a level easily handled by every member of the field party; the leader should never force a group member to do anything he or she is afraid of doing.

Never travel alone; nowhere does the maxim "safety in numbers" hold more true than in the field. Although this rule often goes unheeded, the hiker who travels alone is asking for trouble. The slightest incapacitating accident—a sprained ankle or a twisted knee—and the lone fieldworker is in grave danger. If an accident occurs in a party of two, the uninjured person can go for help. In a party of three, one person can go for help while one stays with the injured person, or both can carry the injured person out. Obviously, the more challenging the terrain and environment and the longer the trip, the more important this rule becomes.

Don't wear yourself out. Working or hiking to the point of exhaustion will impair both your mental judgment and physical abilities, greatly increasing your risk of accidental injury. If you are extremely fatigued, recuperate before taking on any strenuous activity.

PLANNING FOR FIELD SAFETY

Landslides and rockfalls are almost everyday occurrences in some areas. Avoid potential slide areas, and watch for falling pebbles, which often precede rockfalls.

When in the field, stay out of caves and abandoned mines.

Avoiding Falls

Falling is among the greatest dangers in remote areas. Injuries from falling are often severe and can be greatly compounded if the victim falls into a ravine, down a scree slope, or into some other hard-to-reach place, particularly if the weather poses a serious exposure hazard.

Falling is often caused by loose footing, so avoid hiking on loose rock in areas where a slip could result in serious injury. Be hesitant about crossing steep active scree slopes until you've had a lot of practice on shallow gradient slopes. Beware of sharp drop-offs below scree slopes.

Don't jump or bound from rock to rock, especially when rocks are wet or lichen covered. Avoid wet rock slopes. Take care when crossing log "bridges," and stay off fallen logs, which can be very slippery when wet.

If faced with a difficult rock climb, seek an alternate route; mountaineering in the pursuit of research should be avoided unless it is absolutely necessary and thoroughly planned. Never mountaineer unless you and your partner have been properly trained. In addition to the danger of falling, mountaineering is very time consum-

SAFETY IN THE FIELD

ing, and the extra equipment it requires can hamper efficient fieldwork.

Unbelayed climbing of vertical exposures is risky anywhere; in remote areas it is an irresponsible practice. Before taking any risks, consider the potential consequences. Avoid taking even the minor chances you might take nearer home or hospital.

When climbing, be sure of your return route; it is usually more difficult than the ascent.

Crossing Streams

Many avoidable accidents occur during stream crossings simply because hikers fail to take the time to find the safest place to cross. Plan routes to avoid dangerous crossings. If there's any question about the safety of a particular crossing, do not attempt it. Find an alternate route, or return during a drier season.

Use extreme caution when fording streams. Avoid strong currents. Don't cross a raging stream. Remember that streams that were easy to cross in the morning may become impassable after a storm or, if snow or ice fed, after a day of sun. If you are wearing a pack, undo the hip belt so you can discard the pack quickly in case you fall.

If you ford upstream from a bend, you stand a better chance of being carried ashore if you slip. But use caution: the outside edge of bends may be undercut, resulting in entrapment. One recommended way to cross a stream is to have two or more people link arms and

PLANNING FOR FIELD SAFETY

walk in an out-of-step fashion angling slightly down current. The mutual support provides stability.

Ropes can be dangerous unless used with care. When crossing a sizable stream, tie a rope around your waist, and have someone onshore pay it out as you cross. Use a walking stick on the downstream side for added stability. Wear your boots (without socks) to avoid injuring your toes or losing your balance. Don't wear waders on deep or difficult crossings: if you fall and they fill with water, you may not get out; if you fall and they don't fill with water, air trapped inside may hold you upside down.

When crossing a stream on a log, unfasten the waistband on your pack so that you can get the pack off quickly if you fall.

If you fall while crossing a stream, try to get into a sitting position facing downstream with your feet held near the surface to absorb impacts with river obstacles and to keep feet from getting trapped in rocks on the river bottom. This position lets you maneuver to shore while reducing the risk of serious head injury. In swiftly flowing streams, don't try to stand up in water over your knees; if a foot gets wedged under a rock, you could drown. Swim out of rapids. Avoid strainers—partially submerged trees or vegetation that allows water to flow through but which will trap a solid object like your body. If unable to avoid a strainer, swim toward it and leap onto it to get your torso out of the water.

Slow, meandering streams present different hazards, such as muddy water that can obscure sudden changes

SAFETY IN THE FIELD

in depth. Use a walking stick to probe for sudden drop-offs. If the water is deep, let the shortest person lead the way; if the route is passable for that person, it should be passable for all.

Beware of thin ice on ponds. Look for an alternative route around the pond. If you do start to break through, fall flat, kick your legs to the surface, and crawl to safety ice, keeping your weight spread out over as wide an area as possible.

Snow- or glacial-fed streams are colder than they look and can cause muscle cramps. Death can occur in as little as fifteen minutes in water at 32°F. In very cold weather, do not ford streams deeper than knee-deep, and plan your routes accordingly. You'd probably make it across deeper streams, but instantaneous numbness and hypothermia could result and prevent you from hiking farther. Always wear boots or alternative shoes when crossing cold streams. Feet become instantly numb in the frigid water, and it is possible to suffer severe lacerations of the feet without knowing it. Take off socks to keep them dry.

Never wade through streams on top of glaciers. It is virtually impossible to regain lost footing in a rapidly flowing supraglacial stream, which often disappears below the ice downstream.

Quicksand

A gruel-like slurry of loose sediments and ground water, quicksand is most often found in swampy areas and around streams where the water table is high. Although stepping into quicksand can be alarming—each tug on one leg sends the other deeper into the mire—it is not truly dangerous if you don't panic. If you can float on water, you can easily float on the much denser quicksand.

If you get stuck in quicksand, remain calm. Slowly lie back with your arms spread out to distribute your weight as widely as possible. Gradually work your legs to the surface. When your legs are free, roll toward solid ground. Alternatively, a friend can throw you a tow rope or lay a log across the quicksand—taking care, of course, not to get mired in the quicksand.

If You Get Lost . . .

Stop. Try to get your bearings. Wandering around aimlessly will only worsen your situation. If possible, triangulate with your map and compass to pinpoint your position. If you're unable to determine your location and can't figure out which way to go, stay put and signal for help. Three of any sort of signal—for example, three fires in a row, three gunshots, three whistle blasts—is considered a distress call. Don't panic. You're carrying that emergency equipment for a reason; now's the time to use it.

In general, it's unwise to travel at night. If you get lost after dark, stay put, even if you think you know which way to go.

If a search plane comes looking for you, use whatever means are available to attract its attention—lighting fires, waving your signal cloth, using a signal-flare pen, flashing a signal mirror. If you light fires on a sunny day, be sure to burn some green boughs to create smoke and increase visibility. Smoke from a fire or a smoke bomb is a most effective way for a pilot to pinpoint your location. Do not use a red flare unless you are in distress. A red flare will initiate a search-and-rescue operation by the U.S. Air Force or other search-and-rescue organizations.

Once the pilot spots you, indicate your proposed direction of travel by motioning toward it with both arms, or sit down on the ground and motion downward to show that you're staying put. The pilot should wag the plane's wings to show that he or she understands.

Food and Water

The field is not the place to go on a diet. Try to maintain a substantial, well-balanced diet. Fatigue, dizziness, weakness, or stress resulting from an unbalanced or insufficient diet will make you more likely to have a serious accident, may impair your judgment in a crisis, and may make it more difficult for your body to function properly (e.g., to fight hypothermia). On long days, stop frequently to eat and drink. Food consumption in the field can be twice that at home, and even higher in cold weather. Always have enough food for your planned trip, plus an emergency supply.

PLANNING FOR FIELD SAFETY

Contaminated Food

Food that is improperly stored or prepared or meat from infected animals can pose serious health threats if ingested by humans. Wash all fruits and vegetables in treated water before eating. Store reconstituted dehydrated food as carefully as you would store fresh food. Never eat spoiled meat. If there's any chance that game might be infected with a parasite, cook it thoroughly.

Some of the more common illnesses posed by food are discussed in this section. Consult a physician if you think you may have contracted one of the illnesses.

Trichinosis. Trichinosis is caused by a parasite that is most commonly found in such game as bears, pigs, and some marine mammals, and can be ingested by humans if the meat is not cooked sufficiently. Early symptoms include nausea, diarrhea, and intestinal cramping; later symptoms include muscle soreness, eye pain, fever, and other ailments that can last up to three months. Occasionally, trichinosis is fatal.

Alveolar hydatid. This disease, usually fatal in humans, is caused by a particular species of tapeworm (*Echinococcus multilocularis*) found in foxes, dogs, and cats. Various rodents are the intermediate host. When the eggs are ingested by humans, either by drinking contaminated water or eating contaminated food, the eggs release larvae that find their way to the liver. There they proliferate by budding, destroy the surrounding hepatic tissue, and commonly spread to adjacent organs, the lungs, and the brain. The disease occurs worldwide in northern latitudes and is endemic to St. Lawrence Island

and western Alaska, particularly where foxes are abundant. The only adequate treatment is prevention. When near native villages, camps, or fox denning areas, use only boiled water (chemical treatment doesn't help) and clean or well-cooked food. Wash your hands before heating or handling food, and keep nonfood items, such as fingers and pencils, out of your mouth. Obviously, avoid fecal matter, and do not eat any vegetation.

Rabies. Humans can contract rabies not only by being bitten by an infected animal, but by eating infected meat without cooking it sufficiently. Although rabies is most common in foxes, skunks, coyotes, wildcats, and some species of bats, meat from any carnivorous animal is a potential source of infection. Eating infected meat is particularly insidious because, unlike being bitten, the victim has no reason to suspect he or she has rabies—and by the time symptoms appear, it is usually too late to save the victim.

Contaminated Water

Natural water found in the field today cannot be considered safe to drink without first treating it to remove contaminants. Untreated, even the cleanest looking mountain stream can contain a wide variety of viruses, bacteria, amoebas, and protozoa, many of which are capable of causing illness.

Some of the more common illnesses posed by untreated water are described in this section. If, no matter how careful you are, you should contract one of the illnesses, consult a physician for treatment.

Giardiasis. A protozoan parasite whose cysts are spread by human and animal waste, *Giardia lamblia* is found in nearly all water of the American West and throughout other parts of the United States and the world as well. If a sufficient number of these cysts are ingested, susceptible people (about half the population) may, after a short incubation period, contract giardiasis. The illness is marked first by fever, chills, headache, vomiting, and fatigue, and later by explosive diarrhea, cramps, and weight loss lasting a week or more. Sometimes giardiasis becomes a recurring condition that must be treated with powerful drugs.

Hepatitis. The hepatitis virus, particularly hepatitis A, is common in polluted waterways throughout the developing world. Its symptoms range from mild flu-like illness to fatal liver failure. Hepatitis B and non-A, non-B hepatitis, transmitted by untreated water, contaminated food, or body fluids, are less common but more serious; each can lead to chronic hepatitis, which will greatly shorten a person's lifespan.

Schistosomiasis. Schistosomiasis, also called bilharziasis, is caused by flukes whose larvae, released into streams by snails, are able to penetrate human skin and cause weight loss, weakness, headache, nausea, muscle pain, diarrhea, and sometimes chronic diseases of the liver, lungs, intestines, or bladder. Symptoms can take years to develop and often lead to an early death. Schistosome larvae are common in sluggish, shaded streams and quiet waters of South America, Asia, Africa, and the Caribbean. If you're in an endemic area and there's any question that the water may contain this larvae, don't wade, swim, or bathe in it—and certainly don't drink it without treating it.

Alveolar hydatid. A usually fatal disease in humans, alveolar hydatid can be contracted from contaminated water or food. See "Contaminated Food."

Other illnesses. Drinking contaminated water can also cause dysentery, marked by severe, bloody diarrhea, and other intestinal disorders.

Water treatment. There are three main ways to treat water in the field: boil it, treat it chemically, or strain it through a filter.

Boiling is not the most convenient method, but it certainly is the most reliable. Water boiled for at least 10 minutes at sea level (add a minute for each 1,000 feet of elevation) should be free of all potentially harmful organisms.

A number of chemical treatment kits for purifying water in the field are available from camping supply stores. Many involve the use of iodine-based tablets or crystals. One type of kit has tablets containing tetraglycine hydroperiodide. Many camping supply stores also carry water purification kits that contain pure chlorine tablets to sterilize the water and concentrated hydrogen peroxide to bubble off the chlorine gas. Chlorine-based halazone tablets, once a popular treatment method, have proven less stable and effective in the field than iodine-based tablets and are generally no longer used. More convenient than boiling, chemical treatment, if done properly, should remove most harmful organisms. Note that chemical treatment will not remove alveolar hydatid. Be sure to read instructions before using these kits; remember to allow longer treatment time for cold water. Be sure to keep bottles of iodine tablets tightly capped; prolonged exposure to air or light will reduce the tablets' effectiveness.

If you plan to boil or treat your water chemically, it's a good idea to bring an extra set of water bottles; one set can be used for water that's cooling or reacting with iodine, the other for drinking.

Filtration is the most convenient method of water treatment and also the most expensive. The better filters will reportedly remove most contaminants except viruses. If you'll be working in a country where the hepatitis virus is a problem, stick to the other methods.

In Camp

Obtain any necessary permissions and permits before setting up camp. If you're camping on public land, be sure to stay within prescribed camping areas.

When locating your campsite, know the prevailing wind direction, and situate your campsite to diminish its effects.

Avoid wet ground. In dry times, stay clear of low-lying areas that look like they've been flooded in the past. Never pitch camp or park a vehicle on a dry river bed; one good rain and your campsite could be washed away. Always keep an eye out for electrical storms and flash floods and prepare accordingly. Pick a campsite that will be least likely to be a target for lightning, for example, away from the crest of a ridge.

Try to camp near a reliable source of safe drinking water.

Pick a campsite away from avalanche or rockslide danger. Don't pitch a tent under a dead tree or overhanging branches that could come down in a strong wind.

Don't tempt bears or other animals into your campsite with carelessly kept food or garbage. Store food and garbage in plastic animal-proof containers or hang them between trees at least 12 feet above the ground. (In some national parks, anyone going into backcountry is required to use such containers.) Don't eat in your tent or sleeping bag or store food in your tent, pack, or pockets. Vermin will eat through a pack or tent to get to food just as readily as bears will. Do not sleep in the same clothes you cook in. You should pitch your tent where smoke and food odors cannot permeate the tent material.

Carry all refuse out. In rare cases where that is impossible, burn your refuse and bury the residue. Sanitary napkins and tampons should be buried well away from camp or carried out.

Don't litter. Keep orderly latrine habits. Remember that someone may use your campsite after you. Few things are worse than arriving at a "wilderness" campsite that's littered with broken beer bottles, empty sardine tins, and drifting toilet paper. Respect the wilderness; leave nothing but footprints.

PLANNING FOR FIELD SAFETY

Never dig a latrine within 50 yards of a stream, pond, lake, or any body of water.

Be security conscious. If your campsite is near a road or a heavily traveled trail, think twice before leaving it unattended; you may want to hide all valuables or carry them with you.

Campfires

In choosing a site for your campfire, select an area away from anything that can burn or melt, such as tents, sleeping bags, or vehicles. Clear the area to bare ground or rock, making sure there is no root, peat moss, or other organic matter that could ignite and cause the fire to spread. It's a good idea to line the site with rocks, but don't pull rocks from a stream; the more porous rocks are apt to be full of water and explode when heated.

If possible, remove the sod from a fire pit so that you can replace it when you break camp. In a few days, no one will ever know you camped there.

Never leave a campfire unattended. A fire burning out of control can consume a tent site in minutes, leaving you unprotected from the elements even if you escape injury. In the presence of any fire or flame, show constant vigilance, particularly in cold, wet weather, when your guard is likely to be down.

Keep water nearby in case your campfire does get out of control.

Before abandoning your fire, thoroughly douse it with water, stirring through it to make sure no ember remains

SAFETY IN THE FIELD

unquenched. Bury the ashes and return any rocks used to line the fire site to their former locations.

Wildfires

Wildfires, whether caused by human activity or natural processes, are a common occurrence in many areas during dry seasons. Fieldworkers in such areas should be extremely careful not to accidentally touch off a blaze.

Find out what the fire danger rating is for your area before heading out. If it is high, carry a small fire extinguisher capable of putting out brush and flammable liquid fires. Carry a shovel or ax in your field vehicle. Don't build a campfire; use a portable stove for cooking. Watch for sparks when using electrical equipment.

Don't allow heat sources—such as truck mufflers and hot stoves—to come in contact with dry tinder.

If you smoke, do so with extreme caution. Never smoke while hiking; falling ashes could start a fire without your realizing it. Never smoke around flammable liquids or inside a tent, sleeping bag, or other shelter whose fabric could burn or melt.

See "Stoves" in chapter 2 and "Campfires" for specific safety precautions regarding the use of controlled fires.

Keep an eye on the horizon for columns of smoke that could indicate an approaching wildfire.

If you find yourself downwind of a fire, evacuate the area immediately. Remember that the fire will move in the direction of the wind and in an uphill direction, when selecting your evacuation route. Don't wait until fire

PLANNING FOR FIELD SAFETY

surrounds you; your chances of escape are small. Although some people have survived encroaching wildfires by lying in shallow foxholes and covering themselves with soil, try not to let yourself fall into a situation where such last-ditch efforts become necessary.

4

TRANSPORTATION

On and Off the Road

Most accidents on field trips occur while fieldworkers are driving to and from work sites. Use good judgment when driving in the backcountry. Never take a vehicle where it isn't designed to go. Stay alert; don't drive if you're exhausted or under the influence of alcohol, medications, or other drugs. Always follow safe driving procedures, whether on- or off-road, such as maintaining proper distance behind other vehicles and traveling at safe and legal speeds.

Never travel alone. Leave an itinerary with someone who is in a position to help if you don't return at an appointed time. For very remote areas, try to arrange radio contact with a base station.

Pack your vehicle so that all items are strapped down securely. Loose cargo can become dangerous projectiles during sudden stops. Be aware that incessant bouncing may damage or destroy equipment and samples.

Distribute some extra sets of keys among members of the field party.

Carry proper automobile insurance. Double-check your deductible, and be sure you can pay it if necessary.

PLANNING FOR FIELD SAFETY

See "Fieldwork in Other Countries" for information on driving outside the United States.

Choosing a vehicle. Rent an enclosed vehicle or a pickup truck with a covered bed, if possible, to provide a place for people to sleep in inclement weather and to protect belongings and equipment from theft or animals.

If working in remote, difficult terrain, try to rent a vehicle equipped with a winch, a handy device for freeing stuck vehicles and for many other purposes.

Some vehicles come equipped with a handcrank that can be used to start the vehicle in the event the battery dies (it also comes in handy during tuning). Make sure this option is included with your vehicle, if available.

Make sure you'll be driving a vehicle for which spare parts are available in your region.

Parts and equipment. Always have two spare tires in the vehicle, preferably snow tires or large sand tires, depending on the region. Vehicles are usually supplied with only one spare tire; an additional one may be requested from a rental company. Some vehicles are equipped with undersized spares intended for short trips only. This type of spare is unsuitable for rough terrain; if possible, exchange such spares for the regular type.

The person renting a vehicle is usually responsible for replacing damaged tires. Nonradial tires cost considerably less to replace than radial tires do. You may request that the vehicle be equipped with nonradial tires. Bear in mind, however, that their off-road performance is apt to be inferior in snow, ice, and sand.

TRANSPORTATION

Be sure that the vehicle has a functioning jack, as well as a cross lug wrench. These items are not always included with the standard equipment. A can of penetrating oil (WD-40) can make a tire change much easier.

If you need to change a tire, set the hand brake, and block the wheels before jacking up the vehicle. Never work under a vehicle supported by the jack only; lower the vehicle first onto logs or other sturdy supports.

If your vehicle is not equipped with a winch, bring along 10 to 20 feet of chain or a towrope. Some jacks (e.g., Handiman) can be used as a winch. If you need to use a winch, be careful: cables sometimes snap under tension, becoming lethal whips. Draping a blanket or tarp over the cable can reduce this hazard, but one must take care not to let the tarp get caught in the take-up spool.

In areas of few trees, such as desert or tundra, carry a stout steel bar to use as an anchor for your winch.

Always carry more gas than you think you'll need. Using a vehicle with a double tank can increase the driving distance between refueling. If your projected driving distance approaches the limits of the vehicle's tank, carry at least two extra 5-gallon cans, along with a funnel with a filter in it. Off-road or four-wheel drive travel consumes fuel much more rapidly than highway driving. Also carry a small hose and squeeze pump for siphoning in emergencies. Walking two days for gas is rather an inefficient use of field time. Check the gas cans for leaks before heading out; leaking gas or gas fumes are extremely dangerous.

PLANNING FOR FIELD SAFETY

Carry a fire extinguisher capable of putting out gasoline fires.

If you'll be driving far from civilization, you should carry a complete change of oil, an oil filter, a distributor cap, a new set of points, a condenser, and perhaps spare leaves for the springs and a spare axle (commonly supplied with Land Rovers). A spare oil filter and air filter are also advised if you will be driving in very dusty terrain for long periods.

Sand tracks should be carried to provide emergency traction in sandy desert or beach areas. Bring tire chains for ice, snow, and muddy conditions.

If there's a chance you'll be driving at night, make sure the headlights work before you leave. Carry a spare fan belt, jumper cables, extra oil, brake fluid, and coolant. Emergency tools should include wire, hose tape, vise grips (locking pliers), a 10-inch adjustable crescent wrench, a medium-sized regular screwdriver, a medium-sized Phillips screwdriver, a thin file, a flashlight, and flares. Try to find a basic repair manual.

You should be responsible for the general maintenance of your vehicle. Know how to check the oil, brake fluid, battery water, power-steering fluid, and radiator coolant. Know how to engage your vehicle's four-wheel-drive mode.

A shovel or spade should always be in the vehicle. If you are in forested terrain, an ax or large saw is a must for clearing roadways. An ax, shovel, and bucket may be required in some national forests, for use in fire fighting. In a tropical jungle or any area with thick undergrowth, a machete is essential.

In hot or arid regions, carry enough drinking water in the event mechanical breakdown forces you to walk back to civilization. Be sure to carry a supply of portable water bottles.

All vehicles should be equipped with a well-stocked medical kit. It should include essential items as well as items that are too cumbersome to carry on traverse, making the vehicle a more elaborate first aid base than your backpack. At the beginning of each excursion, think about the consequences of having either a personal injury or a vehicle breakdown in the most remote site you will visit, and plan a response to such difficulties.

Driving defensively. Always fasten your seat and shoulder belts. Many states require seat belts, so learn and follow the local laws.

Watch the weather in your area and throughout a drainage basin. Roads on expandable clay can become impassable when wet. Dry canyons can rapidly become raging torrents due to a thunderstorm upstream. Use extra caution on one-lane roads, particularly when driving on blind curves and going over hills.

Some fieldworkers have been killed in collisions with wild animals and open-range animals. On open range you will have to pay for animals you injure or kill. Night travel requires extra caution.

If you will be using active logging roads, contact the logging company to determine periods and routes of maximum use. If you have a radio, maintain contact with moving trucks to let them know that you're on the road. It's your responsibility to get out of the way of loaded logging trucks. The trucks may drive on the wrong side of the road when rounding curves and cannot stop or change direction quickly.

If you're working near a town or city, watch out for three- or four-wheel all-terrain vehicles careening around the hillsides. A slow-moving vehicle of fieldworkers may catch them by surprise.

Remember that you're not driving a sports car or a low-rider. Most off-road vehicles have a high center of gravity, and rolling them on tilted road beds, scree slopes, and logs is a dangerous possibility. Exercise caution. It is generally not a good idea to drive parallel to steep slopes. On tight, steep routes, remember that you may have to back your way out, so make sure this is possible before proceeding.

If you'll be driving an all-terrain vehicle, trail bike, or snowmobile, learn to ride it safely before taking it into the field. These vehicles are inherently unstable and are dangerous when operated improperly. Use good judgment. Always wear a helmet and eye protection.

Off-road driving. Following back roads can be a challenge. In some regions they consist of an interwoven set of tracks a couple of miles wide that all wander to the same place. Often a set of tracks leads off the main road to a dead end. Off-road guidebooks may be available in some regions and are worth using. Local rangers, landowners, logging personnel, and geologists may be good sources of information on back-road availability and conditions. Tracks that appear on 20-year-old topographic maps may have long since been swallowed by the wilderness.

To avoid damaging fragile vegetation, never drive off the road unnecessarily. Tracks can last hundreds of years in some desert, tundra, and alpine meadow areas, and in some pasture land as well.

Drive slowly through streams and mudholes; you'll be less likely to splash water onto your vehicle's coil, which causes the engine to stall, and you'll avoid damage from hidden rocks and potholes. After fording deep streams, be sure to test your brakes—they are usually ineffective when wet. If you're stalled in a stream, you might be able to get out on the starter motor alone. Be sure to remove the spark plugs first to reduce engine compression.

Don't drive in a dry stream bed unless you're sure there's no chance of a flash flood.

Don't park in tall, dry grass. Your muffler or catalytic converter may be hot enough to start a fire. In some dry areas, spark arresters are required on your exhaust system before you can travel off-road.

Snow Vehicles

If you plan to use a snowmobile, snowcat, or other snow vehicle, learn how to operate the vehicle safely, and practice riding it before heading into the backcountry. Learn and follow all laws pertaining to the safe use and operation of your vehicle.

PLANNING FOR FIELD SAFETY

Be alert to the dangers of wind chill when riding a snow vehicle. If you'll be traveling alone into a remote area (not recommended), be prepared to camp overnight if your vehicle breaks down.

Avoid avalanche areas.

If your route will take you across a frozen lake or stream, check with local snowmobilers to make sure it's safe. Before crossing, double-check the ice conditions yourself.

Keep a constant eye open for such obstacles as wire fences and boulders that may be hidden beneath the snow.

Learn general maintenance for your vehicle, and carry necessary repair tools.

Pack Animals

Remember that horses, mules, and other pack animals are living creatures that suffer from heat, cold, hunger, thirst, and fatigue, just as you do. Make sure loads are evenly balanced, and never overload or overwork your pack animal. Keep a close eye on the animal's condition at all times. Do not take pack animals on a field project unless you are qualified to do so.

If you'll be taking a pack animal on an extended field trip, it's essential that you first learn how to care for and ride it properly, even if you'll have a professional handler with you.

The handler or you should check shoes on horses and mules.

TRANSPORTATION

Horses should never be allowed to gorge on grain. Grain should be fed in small controlled amounts determined by the handler.

Learn how to saddle your horse properly. Be careful not to give it saddle or cinch sores.

Never leave horses tied to trees at night. Horses will chew the bark off a tree, possibly killing the tree. Horses tend to paw the ground when tied short, leaving a very unsightly mess in the morning, so make sure your rope is long enough. Use a 30-foot rope, usually the lash rope, to stake out horses. They must be trained not to fight the rope in the event they become entangled. Horses may be hobbled along with picketing. By staking out or picketing horses, they can graze all night. Satisfied animals are not as ready to leave camp as those who are hungry. Do not allow horses to graze down an area and destroy the grass. Move them often.

Never wear sneakers or other shoes without heels; these can slip through the stirrups, trapping the foot.

Don't take your horse into hazardous areas. If you have any question about the terrain ahead, dismount and inspect on foot. Lead your horse through dangerous sections. Many horses don't like to walk through dense brush or other terrain where they can't see where they're placing their feet.

To mount a horse, stand to the horse's left (almost all are trained to be mounted from this side) facing its head, with the reins held in your left hand. With the same hand, grab onto the horse's mane or the saddle horn. Observe the horse's ears; often if they are back and down it means

PLANNING FOR FIELD SAFETY

the horse is agitated and may be ready to bolt. Horses can also bite or kick severely when you are trying to mount. If the ears are up and the horse seems calm, twist the stirrup half a turn clockwise, place your left foot in the stirrup, and swing your right leg up and over the horse.

Most western saddle horses are trained to respond to rider instructions as follows:

To go forward, nudge the horse in the ribs with your heels.

To turn left, hold the reins evenly in your hand and move your hand to the left. This motion causes the right rein to fall across the right side of the horse's neck, signaling the horse to turn left. Note that tugging on the left rein, a technique employed by many neophytes, is not the proper signal and can cause the horse pain and injury.

To turn right, reverse the above.

To stop, pull back on the reins evenly.

To move backward, stand in the stirrups and pull gently back on the reins. Be careful not to tug too hard: the horse may rear and dump you.

When leading a pack animal, never tie the lead rope to your saddle. If the pack horse goes down, it can pull your horse, along with you, down also. Don't allow the lead rope to get under the tail of your saddle horse. This is very irritating to horses. When leading more than one pack horse, tie the horses in such a way to allow them to break apart if one goes down. Use one-quarter inch rope looped through the pack saddle tree; then tie the lead rope through the loop formed by the one-quarter inch rope. Never tie head to tail.

Airplanes

If you will be using aircraft extensively, enroll in a private pilot ground school, and take basic flying lessons. Knowledge of the pilot's workload and responsibility will increase your respect for him or her as part of your team.

Try to learn to read weather and terrain with the eye of a pilot. Ask your pilot to explain the types of conditions that are best for him or her. If you'll be landing on bodies of water, learn to read the warning signs of submerged dangers.

Learn beaching and docking procedures so you can help the pilot if your help is requested during an emergency.

Remember that a float plane behaves just like a boat after it has landed and that boating safety instructions apply. Read "Safety on the Water." If using a float plane, know how to swim.

Before a flight. Before leaving the aircraft base, the pilot will have filed a flight plan based on your intended mission objectives. You should make sure that someone remaining behind has a copy of your itinerary, including all flying routes and the location of your camp. Mark this location on a map so there won't be any mistake. If you're not sure about the accessibility of your first campsite choice, mark your second choice on the map

PLANNING FOR FIELD SAFETY

as well and clearly label it. Then, the pilot only has to alert others to the fact that you're at the second site, without having to remember to mark it on the map. Describe what you will do if you are not picked up as planned. This precaution is important.

Before your flight, plan and clearly mark your route on a large-scale topographic map. Try to choose routes that avoid downdrafts, which are common near big mountains and valleys, because they place severe demands on aircraft and pilot. Review your route and project objectives with the pilot to ensure that the pilot and the aircraft can complete the intended mission.

Be prepared to navigate as you go, a task that often proves difficult for the inexperienced. The easiest method is to begin tracing your route as soon as you leave the ground, taking care not to lose track of your position during the flight. If you plan to look at geology from the air, mark the areas on the map also. Once in flight, there is no time for planning.

Know the volume and weight of all personnel and equipment so the pilot can determine aircraft suitability. If the pilot objects to the amount of your equipment, leave nonessentials behind or make two trips. Be sure to discuss your plans with the pilot well before your flight to prevent hasty, last-minute decisions. If you will be returning with more weight than you left with (e.g., a large number of rock samples), discuss your plan with the pilot before leaving so he or she can plan fuel weight accordingly.

TRANSPORTATION

The pilot should load the plane or at least supervise the loading; he or she is the one who will have to compensate in flight for an uneven load. Load heaviest gear first onto the floor of the plane. Make sure the load is balanced between forward and aft sections. Don't overload the plane. Make sure the load is securely stowed or tied down to prevent shifting in flight.

Pack flammable and explosive gear, such as fuel, flares, and dynamite, separately and within reach. It may be necessary to throw it from the plane prior to a forced or crash landing. Be sure to inform the pilot of any hazardous cargo. Never put a loaded firearm in an aircraft.

Respect the pilot's judgment on weather conditions for flying, suitability of landing sites, and amount of gear that can be carried. Never try to coerce or harass a pilot to act against his or her judgment. The pilot's confidence in you may grow as he or she becomes more familiar with your mission. Never fly with a pilot in "bush conditions" unless the pilot is qualified; always check references.

Follow the pilot's instructions at all times. The pilot is in command of the aircraft and responsible for safe operations and for you as long as you are in his or her care. The pilot should brief passengers on safety procedures, which must be followed. Do not hesitate to ask questions if any instructions are unclear to you.

Try to arrange your flight as early in the day as possible. An early flight will give you more time to get

PLANNING FOR FIELD SAFETY

to your destination and set up camp before nightfall and will help you avoid afternoon winds and storms.

Always dress appropriately when flying over remote country. Even if you don't plan to get out and walk, you may be forced to if there is an accident. Hypothermia can kill as effectively as a crash. Be sure to store any extra clothing where you can get it easily, not locked in a cargo compartment.

Have extra, dry clothes readily available if you're being dropped off by a float plane; you will be required to wade to shore.

If you'll be using aircraft to ferry groups of fieldworkers to a base camp, make sure that each planeload carries enough food, sleeping bags, and shelter for each passenger aboard in case all trips are not completed that day, or in case the aircraft must make a forced landing in the wilderness. Carry firearms if appropriate.

During a flight. Never distract the pilot during flight. However, if you notice an approaching aircraft or other danger, make sure the pilot is aware of it also.

Sit in the seat assigned to you by the pilot for weight and balance purposes. If you're sitting in the copilot's seat, keep your feet and hands clear of controls. Secure small loose items that might be thrown around in rough air.

Arrange with the pilot when and where you want to be picked up. Ask the pilot to fly to the pick-up site before dropping you off to make sure that the landing

TRANSPORTATION

site is suitable. Arrange an alternate pick-up site in case the first becomes unusable.

Whenever possible, fly along your traverse route to make sure you can cross it—no major rivers or cliffs. Do not change your prearranged pick-up locations after you are airborne unless you can be assured that you can maintain continuous radio contact with the base.

On approach to the landing site, indicate the general area where you want to land, but let the pilot pick the exact landing site. It is his or her decision. Don't expect to be able to taxi over to the spot where you want to start a traverse; the pilot will try to land as close to that spot as conditions permit. What may be an acceptable landing site to one pilot may be beyond the capabilities of another or his or her aircraft.

Pick-up site. Bad weather or mechanical difficulties regularly ground aircraft, so don't count on being picked up on the specified day. Carry enough food, clothing, and emergency gear to survive in the field for several extra days.

If you are not picked up on the specified day, it is generally best to stay where you are and wait for a later pickup. Don't walk or hitch a ride out, unless you have made radio contact with a base camp and the airbase. Many costly and dangerous searches have been made looking for someone who hitched a ride home and didn't tell anyone.

If you're being picked up by a float plane, do not enter the docking area until the aircraft has come to a complete stop. If you are assisting in the docking process, crouch

PLANNING FOR FIELD SAFETY

down on the dock as the plane approaches to prevent a wing or propeller from knocking you into the water, and perhaps unconscious.

Be aware that high winds or high hills can make a fairly big lake difficult or impossible to take off from safely. It may be impossible to turn a plane around on a windy lake without capsizing; you may have to taxi to shore and turn the plane around by hand.

Avoid the propeller area completely until well after the propeller has stopped. A "dead" propeller can clobber you with one last kick.

If you crash. Get away from the aircraft as quickly as possible; leaking fuel is highly explosive.

Most aircraft have an emergency locator transmitter, or ELT, that will transmit on a set frequency to summon help. An ELT is supposed to activate automatically when an aircraft crashes, but can be operated manually as well. Make sure your pilot shows you how to operate this radio in case he or she is incapacitated during an accident. Different signals can be sent to rescuers to indicate what happened, such as whether anyone is dead or critically injured and the urgency of the rescue needs.

Helicopters

Before flying in a helicopter, you should know how to enter and exit the aircraft, how to work the doors, how to fasten the seat belts, and where first aid and survival gear are located.

TRANSPORTATION

Do not approach the helicopter until the pilot signals you to do so. Stay in view of the pilot as you walk toward or away from the helicopter.

As you approach or leave a helicopter, keep your head low to avoid having it struck by the rotor blades. On windy days or on uneven terrain, the blades can come as close to the ground as four or five feet or less, so you should always walk in a crouched position near the helicopter. Do not walk uphill from the helicopter. Do not walk directly downhill either. Doing so may interfere with the helicopter's "ground effect." In rare cases where space is tight at the landing site (e.g., on a cliff-bounded ledge) and it is impossible for you to move out of the range of the rotor blades, the safest place to crouch during takeoff is right at the foot of the skids within sight of the pilot.

Never walk behind the helicopter. Injury or decapitation can occur from the tail rotor, which is hard to see when moving. Delicate radio antennas, on which your survival may depend, are commonly located in the tail section. If you must go around to the other side of the machine, always cross in front in view of the pilot.

Don't walk under the tail boom; hair caught in the drive shaft can result in scalping, and exhaust heat can result in severe burns.

You should not smoke in or around a helicopter. Don't make sparks, especially when refueling. Some helicopter fuels are more explosive than gasoline.

Avoid overloading a helicopter. Remember that maximum load decreases with elevation, and with increasing

PLANNING FOR FIELD SAFETY

temperature or decreasing air pressure as the weather changes. A pilot may not be able to lift a chopper off the ground at 7,000 feet elevation with the same load the aircraft was able to lift at 3,000 feet.

When you exit a helicopter, be sure to latch the door securely. Don't slam it; helicopter doors are fragile and expensive.

Be gentle getting into or out of a helicopter that is resting on uneven ground or on snow. Shift your weight slowly and gradually. A sudden loss or gain of weight might make the helicopter tip or slide.

Sit reasonably still while flying, particularly in small helicopters. The pilot must compensate every time you shift your weight.

Don't bother the pilot during difficult moments, such as lifting off, landing, or in turbulent air conditions. Flying a helicopter isn't easy. But don't hesitate to mention dangers to the pilot that you feel he or she hasn't noticed, such as trees, guy wires, and oncoming large birds. Better safe than sorry.

Wear a seat belt at all times. Wear head phones, if provided; helicopters are noisy and can cause permanent hearing damage. Some companies provide helmets for passengers.

If your helicopter makes a crash landing in a body of water and starts to sink, open the door slightly to allow water into the cabin. This action will help equalize water pressure inside and outside the cabin and enable you to open the door fully. Wait for the rotor blades to stop turning before you swim to the surface.

TRANSPORTATION

If you will need to be picked up from a spot that's difficult to locate from the air, be sure to wear bright clothing and carry a fluorescent tarp, a signal mirror (or use the mirror on your compass), and possibly an emergency strobe light. If you'll be using helicopters on a frequent basis, wear high-visibility, fire-retardant clothing in case of a crash or breakdown in a remote location.

You might consider carrying a pocket-sized signal-flare pen with either red or white flares and smoke grenades (bombs) to attract attention. A red flare is only to be used if you are in distress (i.e., injured) because a report of a red flare will initiate a search-and-rescue effort by the U.S. Air Force or other search-and-rescue organizations. White flares or any color other than red can be used for signaling. If there is more than one group in the field, different color flares can be assigned to different groups. If you are using a flare, be sure that the projectile is aimed away from people and nearly vertical to prevent injuries or fires.

Smoke bombs are more effective to pinpoint your exact location for the pilot. White smoke from a fire is also a good indication of your position. The smoke also provides a helicopter pilot with ground-level wind information needed for a safe landing. See "If You Get Lost" for more information on signaling for help.

If flying over large bodies of cold water, wearing an immersion survival suit is recommended to prolong survival in case of a forced water landing. Helicopter seat cushions are not flotation devices, as they are in commercial planes.

PLANNING FOR FIELD SAFETY

If you plan to use a walkie-talkie to communicate with your helicopter pilot, make sure the system is working before he or she drops you off.

If a helicopter is going to land at your camp, observe the following guidelines:

Make sure all gear is secure. Helicopters create strong winds that can blow gear away and rip tents. Put large rocks over tent pegs to keep the pegs from pulling out.

Make sure there is plenty of room for the helicopter to land. It will need a flat, stable landing area at least 30 to 40 feet square. An area of smooth rock is best; snow is worst. Don't set up your tents on or near the landing area.

If the helicopter must land on snow, lay an object on the snow for scale; otherwise the pilot may have difficulty judging how close the helicopter is to the snow.

Set up a wind indicator for the pilot. You can simply hold up flagging tape or a brightly colored cloth—anything that will clearly indicate to the pilot the strength and direction of the wind. If you will have radio contact with the pilot before the landing, be prepared to provide pertinent weather data.

If the helicopter is landing at night (during an emergency evacuation, for instance), use flashlights to illuminate any dangers the pilot may not see, such as tree branches or power lines.

As the helicopter lands, crouch down within sight of the pilot, look away from the helicopter, and cover your eyes; sand or debris may be blown around by the rotor blades.

TRANSPORTATION

Be sure to wait on the side of the landing pad that the helicopter will be facing when it lands (usually the upwind side). Never wait uphill of the landing pad.

Be extremely careful when carrying equipment near the helicopter. Carry long objects, such as tent poles, rifles, and stadia rods, horizontally so as not to accidentally place them in the path of the blades. Don't toss objects out of the helicopter during unloading; they might get blown up into the blades. Instead, carry them away from the helicopter and lay them down, being careful to put heavy objects on top of light ones. There is nothing worse than having a sleeping bag blown over a cliff as the helicopter flies away.

Safety on the Water

Never go boating alone.

Stay away from tangled weeds, as they can trap small boats and swimmers. Fieldworkers have been caught in weeds and drowned in shallow water a few feet from shore.

If you don't know how to swim, learn before your trip. If you don't swim well, advise others in your group accordingly, and use extreme caution when near the water.

Always wear a life jacket, preferably a brightly colored one, in small open boats and on steep shores. Be sure that the life jacket will keep your head above water, even if you're unconscious.

PLANNING FOR FIELD SAFETY

Keep lifesaving gear close at hand. A throw bag, an emergency heaving line stuffed into a floatable bag, makes a quickly accessible water rescue device. Throw bags are available commercially from white-water canoeing equipment companies.

If boating on extremely cold water, wearing an immersion survival suit is advisable, together with a life vest, regardless of the season.

Use navigation charts, which show water depth. Learn the navigation markers and maritime rules. This information can be obtained from the U.S. Coast Guard and the National Oceanic and Atmospheric Administration.

A fire on the beach is an international distress signal. Make sure you don't make your cook fire in sight of the open water.

Learn how to tie basic knots—clove hitch, bowline, tautline (rolling) hitch, sheet bend, square knot, figure eight—and know when to use them. Basic sailing or climbing books usually have a section on knot tying.

Protect cameras, maps, and field notebooks in waterproof bags. Carry waterproof matches only. In rough water, store your belongings in double or triple plastic bags, in case you take on water. Lash all belongings to your boat if there's a danger they will go overboard.

Be alert to sudden weather changes, especially afternoon storms. Avoid crossing lakes or bays during high winds, particularly if using a canoe or small craft. Make sure you have dry emergency food and gear so you can make camp until the wind diminishes.

TRANSPORTATION

During periods of windy weather, traveling at night, evening, or before dawn may take advantage of calmer water conditions. If you do travel at night, be sure to display the proper running lights and carry a spotlight.

When you're working on the ocean, carry extra fresh water. Do not go out of sight of shore during foggy conditions.

Ballast your boat against rough water (but don't overload it). Observe weight and passenger limits for your craft.

Avoid hazardous passages by using alternate routes or landing sites. If you're on a river, be prepared to portage your boat.

If ice is abundant, travel while wind and tide are low to minimize danger.

If you'll be working on the water during cold weather, read "Cold-water Immersion." U.S. Coast Guard pamphlet CG 473, *A Pocket Guide to Cold Water Survival*, is also recommended.

Motorboats

Before taking to the water, become thoroughly familiar with your boat, its equipment, and its operation. Know its capabilities—speed, ability to handle rough water, bottom clearance.

Know your boat's rate of fuel consumption. Always carry more fuel than you think you will need as fuel consumption rates are much higher in rough water. Check fuel supplies regularly, and shorten your trip if

PLANNING FOR FIELD SAFETY

your craft is consuming fuel rapidly. Reserve more than half your fuel for the return trip because foul weather or changes in wind direction or tidal currents can cause increased consumption rates. Stow fuel safely. Refuel at calm, convenient, and safe places, preferably on shore.

Be sure the boat is equipped with a towline that's at least two-and-one-half boat lengths long. Carry oars and a bailing device, such as an empty can.

All members of the group should be able to operate the boat by themselves, in case of emergency.

Know basic engine repair. Check fuel supply and spark plugs if running becomes rough or starting becomes difficult. Carry an engine repair kit containing spare spark plugs, a spark plug wrench, shear pins, a spare propeller, lubricants, pliers, and a screwdriver. Read the engine operating manual carefully before leaving for the field.

Don't operate your boat at excessive speeds, particularly when traveling on rough water. Keep a sharp eye out for floating or partially submerged objects. Take turns slowly. Avoid hitting rocks or sandbars with the motor or propellers.

Be careful not to damage the boat when landing onshore. Come into the shore slowly and cautiously.

Canoes and Inflatable Boats

When canoeing, carry extra paddles within easy reach; don't bury them all under the load. Always carry a bailing device.

Haul canoes and other small boats out of the water, and secure them carefully, either by tying them down or weighting them down with rocks or water. Be aware that bears may mishandle overturned canoes or inflatable boats, tearing canvas in the process (not a problem with aluminum or fiberglass canoes). Porcupines and other animals may gnaw on the salty handles of canoe paddles, so keep the paddles out of reach if animals might be a problem.

If using an inflatable boat, carry a patch kit, an air pump, and an air-pressure gauge. Check the air pressure in each of the chambers regularly to be sure the boat is sufficiently and evenly inflated. Stay clear of sharp rocks and barnacles.

Exercise extra care when packing inflatable rubber boats so that sharp edges of gear don't wear through the sides of the boat.

Make sure a damaged area is completely dry before attempting to affix a patch.

Tidal Areas

Be alert for strong currents and rapids. These are least dangerous at slack tide. Low slack tide, when all hazards are exposed, is the safest time to explore new areas. Mark on your map the locations of the deepest channels.

Be sure to camp above the high-water mark to avoid having your camp flooded when the tide comes in. If you beach your boat, make sure that it, too, is above the high-water mark to prevent its floating away.

PLANNING FOR FIELD SAFETY

Carry an accurate tide table. Do not leave your boat moored and unattended: on rising tide, the mooring point may get flooded; on falling tide, your boat may be stranded. Take care not to get cut off from your boat by a rising tide.

How you choose to moor your boat overnight varies with the size of the boat, the nature of the shoreline, and the size of the tidal range. Mooring your boat incorrectly can result in any of the following, especially with a large tidal range:

Boat is left stranded by a falling tide. It may settle on a steep slope or sharp rocks not visible at high tide.

Boat is pulled under by a rising tide because not enough slack was left in the mooring or anchor line.

Anchor rope frays and breaks by rubbing on sharp objects not exposed when you moored the boat, and the boat may be swept away or dashed on the shore.

Oceanic Research Vessels

Life aboard a research vessel is a highly structured one that takes some adjustment. Discussed here, in addition to the safety aspects of work at sea (which will no doubt be repeated and expanded upon by the chief scientist before each cruise), are suggestions on appropriate conduct while at sea and tips to make the trip more enjoyable for yourself and others.

A ship at sea is a tightly packed microcosm of society. Usually the crew settles into a highly repetitive routine. For example, the radio operator will always sit in the same chair at the same table, and times for rising and showering are strictly set. To an outsider these routines seem designed to make a monotonous life even more monotonous, but they help reduce the friction associated with tight quarters. Ships also usually have restrictions regarding liquor, gambling, and sexual activity. Although these regulations vary from ship to ship, obey those in effect on your ship, even if you think they are unusually strict. You run the risk of being confined to quarters or put ashore for violations of ship's rules.

A research ship is a hazard to feet. Always wear shoes, because an unexpected roll or pitch of the ship can cause you to lose your footing and injure your foot.

If you have any unusual health problems, make sure you get a medical checkup before the cruise. If you have a chronic condition that requires medication, consult your physician about taking extra medication along. It is advisable to inform some members of your group of your medical problems and medications.

If you are unsure of how to operate a piece of equipment, do not touch it until you have been trained to operate it safely. Most ships require but often do not enforce rules requiring you to wear life jackets and hard hats when handling equipment on deck. Those who choose not to wear this equipment risk getting hit on the head with some piece of equipment and winding up unconscious in the water.

Any piece of gear going over the side has the capability to take you with it. Stand clear of all lines, and never stand in a loop of wire or coil of rope. Such precautions

PLANNING FOR FIELD SAFETY

may seem obvious, but things are often confusing when gear is being deployed or recovered, and the obvious is easily overlooked. The ship's crew is responsible for all over-the-side operations. Stay clear of these activities.

Make sure you know what to do in case of fire or any other emergency. Know the locations of life rafts and life preservers. The ship officers will provide instructions and hold drills. Participate in all drills as if they are real.

Don't waste water. If your ship doesn't have evaporators, take only "ship-showers" (turn off the water while soaping up).

Seasickness. Given the right conditions, anyone will get seasick, but fortunately, seasickness usually passes. Seasickness tablets may help some people get over the initial bout, but they should not be taken for more than three days; any longer and they prolong susceptibility. An alternative to tablets is a patch that you wear behind your ear to prevent seasickness. Check with a pharmacist or physician for more information. As a general rule, if you feel bad, eat a few saltines or other salty crackers. Make sure you drink plenty of water to prevent dehydration caused by vomiting. If you are prone to seasickness, locate the gravimeter and ask to sleep in the nearest bunk room, which has the most stable bed on the ship.

Clothing and supplies. Take clothes appropriate for the latitude you are going to, but don't take anything you really value; shipboard laundry facilities are often hard on clothes. If you are joining a non-American ship, take soap and a towel (they probably won't be provided). Even if you're going to the tropics, take a light sweater to wear in the lab; labs are usually heavily air conditioned to protect electronic equipment from heat, salt, and humidity. Take a long-sleeved shirt for protection from the sun.

It's a good idea to take medication for athlete's foot with you; you're bound to get a mild case. Take an antacid; even the saltiest dogs get queasy from time to time. Take a hat with a visor and two pairs of sunglasses (at least one will go overboard). Don't forget to take books, magazines, and other diversions to while away your spare time (there is usually plenty). In general, travel heavy; if you don't have it when the ship sets sail, you can't get it.

5
WEATHER CAUTIONS

This chapter contains information on weather conditions likely to be encountered by fieldworkers in a wide variety of settings. For less common hazards specific to certain regions, see "Hazards of Specific Regions and Settings."

Cold Weather

Never underestimate the dangers of cold weather. Even moderately cool temperatures, when combined with wet or windy conditions, can be enough to sap your body's heat. When working in cold weather, bring plenty of extra warm clothing and dress in layers.

Eat and drink frequently, even if you don't feel hungry or thirsty, to reduce your risk of fatigue and dehydration; these factors can contribute to hypothermia.

Whenever you're away from camp, carry some emergency equipment.

Clothing

The type and amount of clothing you bring will depend, of course, on the severity of the weather you expect to face, the length of your trip, and the type of work you'll be doing. The following information will help you choose field clothing knowledgeably.

Clothing materials. Today's fieldworker has a greater number of outdoor clothing materials to choose from than ever before. Synthetics are fast replacing natural materials for many applications. For the confused, here's a brief rundown of the pros and cons of the most widely used materials.

Cotton. When dry, cotton insulates well against heat and cold and, if tightly woven, gives protection from wind. When soaking wet, however, cotton becomes worse than useless: the fabric absorbs moisture readily, but does not carry, or "wick," it away from the body. Thus, any moisture absorbed, whether from perspiration or the environment, is held against the skin, quickly chilling the body. If you'll be working in cold or wet conditions, stay away from cotton.

Wool. Wool provides more warmth per unit weight than any other natural fiber. Unlike cotton, wool wicks moisture away from the body and retains more than half of its insulating properties when fully saturated. Some people are allergic to wool.

Silk. Almost as warm as wool by weight, silk also wicks water away from the body and retains its insulating properties when wet. Because of its comfortable feel

WEATHER CAUTIONS

and body-clinging fit, silk is most often made into undergarments.

Polypropylene. A lightweight, synthetic material, polypropylene is nonabsorbent. Used primarily in the manufacture of long underwear and other undergarments, polypro, as it is commonly called, wicks perspiration away from the body six times faster than wool, reducing the risk of evaporative cooling. Because polypro fibers cannot hold water, garments made of the material dry quickly. Polypro is available in various weights. Be aware that chemicals, such as insect repellent, can damage polypro. It also melts easily, so exercise caution around sparks or flame.

Pile. A thick fabric made of polyester coils, pile wicks almost as efficiently as polypro, dries quickly, and, like wool, retains its insulating properties when wet. It is usually made into jackets and shirts. Pile affords little protection against wind, however, and should therefore be worn with a wind-resistant outer garment. It is easily melted by flame or high heat and may be damaged by chemicals. Synthetic fleece and bunting are similar to pile and are often grouped generically under that name.

Nylon. A synthetic fiber, nylon is available in a wide variety of fabric weights and weaves. Often used in outer garments, it provides excellent protection against wind if tightly woven and against rain if waterproofed. Nylon is easily melted by flame or high heat and may be damaged by chemicals.

Gore-Tex. This material is a synthetic fabric perforated with billions of microscopic holes through which molecules of water vapor can pass but drops of rain cannot. Theoretically, this construction makes the fabric waterproof while allowing it to "breathe," so the wearer doesn't end up drenched in sweat. Also wind resistant, Gore-Tex is primarily used in outerwear.

Insulating fill. For use in such items as jackets, sleeping bags, and mittens, insulating fill is usually made of down or synthetic (polyester) insulation.

Pound for pound, nothing beats dry down as an insulator. But if allowed to get wet—for example, if you sweat inside your jacket or sleeping bag—down loses its insulating properties. It also takes a long time to dry, particularly under less-than-ideal conditions. Unless you are willing to exercise the extra care down requires, you're probably better off using one of the synthetic fills or, in the case of jackets and vests, converting to pile.

Synthetic insulation is called fiberfill, and common varieties include Hollofil, Quallofil, PolarGuard, and Thinsulate. Heavier and bulkier than down for a given amount of insulation, fiberfill retains its warmth when wet and dries quickly.

Guidelines for warm dressing. In cold weather, your body's top priority is to keep your vital organs warm by conserving heat in the body core, even if this priority means restricting blood flow to the extremities. Your top priority should therefore be to insulate your trunk from the cold. If you allow your trunk to cool, you cannot expect to keep your hands and feet warm.

Wear a hat. Most of the body's heat is lost through the head.

Dress in layers. Layered clothing not only traps more insulating air between your body and the elements, but allows you to add and remove clothing as necessary to maintain a constant, comfortable temperature. Try to anticipate changing temperatures and activity levels so that you remove layers of clothing before you start to sweat and add layers before you start to cool off. It's much safer and easier to keep yourself warm and dry than to try to get back into that condition after you've become cold and wet.

For cold and wet conditions, many fieldworkers are opting for a basic three-layer system of synthetics: a polypropylene inner layer (long underwear), a middle layer of pile, and a windproof and waterproof outer shell. This system can, of course, be modified to suit your particular needs and wants.

Make sure your outer layer is truly waterproof; check all seams for leakage, and reseal with seam sealer if necessary.

Always be prepared for unexpected adverse conditions; when in doubt, bring more rather than less.

Hypothermia

Failing to abide by the guidelines for staying warm and dry can lead to a condition called hypothermia, the lowering of the body's core temperature. What makes the condition so insidious is that as body temperature drops, blood is drawn into the core from the extremities, including the brain; mental processes quickly become sluggish; and the victim is unable to recognize—may even deny—his or her own symptoms and fails to take lifesaving action. Unconsciousness and even death may follow. For this reason, it is essential that all members of a field party be aware of the early warning signs of hypothermia so that they can recognize the signs in themselves and others.

Signs of hypothermia. For treatment purposes, hypothermia is generally classified as either mild or severe. Mild hypothermia occurs as a person's core temperature drops from normal, about 98.6°F, to around 90°F. As the temperature drops, the hypothermic victim shows increasingly apparent signs of the disorder. The first sign, not surprisingly, is that the person feels chilly—an excellent time to put on more dry clothing and eat some high-calorie food, stopping hypothermia from progressing further. If such action is not taken, the person's core temperature continues to drop. At this point, skin often feels numb, movements become slightly uncoordinated, and shivering begins. If the temperature drop is allowed to continue, muscle incoordination and weakness become more apparent. The person may start to stumble. Signs of mild apathy or confusion appear; for instance, the victim may recognize the need to change a pair of wet pants, but has trouble figuring out that boots must be removed first. The person may show signs of overwhelming exhaustion, and as body temperature approaches 90°F, muscular incoordination becomes apparent. The person stumbles and is unable to use his or her hands effectively. Speech and thought processes may become sluggish.

It is essential that hypothermia be caught and reversed in its early stages. Unless all other members of a field party are hypothermic also, which is unlikely, or the hypothermic victim is separated from the group, group members should have no difficulty recognizing the more severe signs of mild hypothermia. Allowing mild hypothermia to progress into severe hypothermia takes the victim beyond the reach of effective wilderness treatment.

The first sign of severe hypothermia, in which body temperature drops below 90°F, is arbitrarily taken to be when the victim stops shivering. Muscle incoordination and weakness are now so severe that the victim is unable to walk or stand and begins to show signs of severe neurologic impairment. Speech becomes incoherent and actions irrational. A victim at this stage may, for instance, show complete disregard for the cold, leaving his or her jacket open and hat and mittens off. In some cases, severe hypothermia victims experience a sensation of warmth and begin to remove their clothes. As body temperature continues to drop, it is marked by severe muscle rigidity, semiconsciousness, dilation of the pupils, and a heartbeat so weak and breathing so shallow that they become nearly undetectable. For this reason, a hypothermia victim should never be pronounced dead unless he or she has been warmed first. Unconsciousness follows and then, when heartbeat stops altogether, death.

Treatment for mild hypothermia. Treatment of mild hypothermia in the field is relatively simple, but must be decisive and aggressive. Move the person into a tent or other sheltered area, out of the wind, as soon as any sign of the disorder is spotted. Wet clothes should be removed and replaced with warm, dry clothes. To take advantage of the body's own warming mechanisms, the victim should do some vigorous exercise, either by hiking back to the warmth of base camp, if feasible, or by jogging in place or doing "step-ups" onto a rock or log. Because depleted energy stores are usually a contributing factor in hypothermia cases, it is important that the victim be given some easily digested, high-calorie food to eat, such as chocolate or cookies. The victim should also be given plenty of warm liquids to drink, both to counter dehydration and to warm the body internally.

Treatment for severe hypothermia. The above techniques will not help the victim of severe hypothermia and could even be harmful. Vigorous activity is out of the question; even in the earliest stage of severe hypothermia, the victim is too incapacitated to exercise. Putting on dry clothing, while important to prevent further heat loss, will do nothing to reverse the progression of hypothermia since the victim's body will have lost the ability to rewarm itself. Warming the victim externally, such as by stripping him or her and placing in a sleeping bag with another person, is also a bad idea: when the extremities are warmed, once-constricted blood vessels will open and send a rush of cold blood loaded with lactic acid into the heart, very likely sending it into ventricular fibrillation, a condition in which effective pumping action ceases. This situation is almost always fatal, because a chilled heart will not respond to resuscitation techniques.

There is no proven, effective field treatment for severe hypothermia. Your best alternative is evacuating the victim. But evacuation poses problems of its own. In severe hypothermia, the heart is so cold that any jostling of the body, such as carrying a person, can send the heart into ventricular fibrillation.

The best solution in cases of severe hypothermia is to evacuate the victim by helicopter. If this means is not possible, however, and a medical facility is less than four hours away, you should evacuate the victim over land. Although the victim's heart will probably go into ventricular fibrillation, the low body temperature and depressed metabolism will probably allow the victim to survive decreased heart action for several hours without permanent damage.

If a medical facility is more than four hours away, you have no choice but to attempt treatment in the field. Place the victim in the warmest environment available. Dress him or her as warmly as possible to prevent further heat loss and to take advantage of the body's remaining metabolic activity. Bearing in mind the importance of warming the central parts of the body before the extremities, you should place external sources of heat, such as water bottles filled with warm water or heated stones covered with fabric, against the chest, abdomen, and the sides of the neck. The arms and legs should be insulated, but not rubbed or actively warmed. If the victim regains consciousness, he or she should be encouraged to drink warm liquids, if it can be done safely. This rewarming process is slow and uncertain; expect it to take longer than 24 hours, if it is successful at all. Bear in mind that the chances of survival for a victim of severe hypothermia are not good, even in a hospital setting.

Clearly, preventing hypothermia, or treating it in its early, mild stages, is far easier than treating severe hypothermia. Be alert to its early signs in yourself and others, and take immediate action to combat it. Never assume that conditions have to be frigid for hypothermia to occur; most cases of hypothermia occur at temperatures above freezing, when people fail to realize the danger of cool, wet weather. Don't let yourself be caught off guard.

Frostbite

Frostbite, the actual freezing of body tissue, is a potential hazard whenever the temperature drops below freezing. Most often affected are the hands and feet, which are far from the heart and can suffer from constricted blood circulation in cold weather, and the ears and face, which are often exposed to the elements.

Prevention. The factors that contribute to hypothermia are among those that contribute to frostbite (the two often occur together), so a prudent first step toward preventing frostbite is to keep yourself warm and dry. If you allow yourself to get chilled, blood vessels in the extremities will constrict in an effort to contain heat in the central parts of the body. Also contributing to poor circulation in the extremities, and thus to frostbite, are tight clothes, particularly boots. Give your feet plenty of room, bearing in mind that feet swell slightly during exercise and felt boot liners tend to expand as they absorb sweat. Cigarette smoking, which constricts blood vessels in the skin, may also contribute to risk of frostbite.

In very cold weather, particularly if the wind is blowing, put on a face mask that covers your ears and nose. Frostbite generally begins with numbness in the affected body part. Sometimes patches of white, frost-nipped skin will appear on a person's face, nose, or other area without the person realizing it. Group members should keep an eye out for these patches on each other's skin.

Be careful handling white gas and other highly volatile fuels in subfreezing temperatures; these fuels can cause instant frostbite if spilled on the skin.

Treatment in the field. A body part in danger of frostbite (numb or frostnipped, but not yet frozen solid) should be immediately warmed by placing it in contact with warm skin, such as under the armpit or against the abdomen. Don't rub the affected area because you can damage frozen skin cells.

Conditions for treatment. If body tissue has frozen solid, however, do not attempt treatment in the field unless the following conditions can be met:

1. You have the proper facilities for rewarming the limb according to the procedure outlined in the next

PLANNING FOR FIELD SAFETY

section. Such facilities include a tub large enough to hold the affected limb and enough warm water to fill the tub and keep it at a prescribed temperature for up to an hour.

2. The patient can be kept warm both during the rewarming process and during the evacuation to a medical facility. If there's any chance that thawed tissue might become refrozen, leave it frozen; refreezing causes permanent tissue damage. Bear in mind also that thawing a frostbitten limb will cause the victim such excruciating pain that, if a foot or leg is involved, the victim may no longer be able to travel under his or her own power. If there's any chance of refreezing or if the victim might need to use the thawed limb before it's completely recovered (which can be several weeks or more, depending on the extent of the frostbite), it's better to leave the limb frozen and numb until the victim can be evacuated to a medical facility.

Procedure. If you're sure the preceding conditions can be met, and distance or other factors prevent quick evacuation to a medical facility, you may decide to treat a frostbite victim in the field. Be aware that there are many myths about the proper way to treat frostbite—pushing the affected body part into the snow, for example, or slowly thawing in cold water—all of them are ineffective, many of them dangerous. An outline of the proper procedure for treating frostbite follows. Consult the first aid manual in your medical kit for more detailed information.

A frostbitten limb should be rapidly thawed in a tub of warm water (100° to 110°F) until the affected tissue is soft and pliable. The limb must be suspended in the water, not allowed to rest on the sides of the tub. When the water temperature drops below 100°F, the limb should be removed from the water until the water's temperature can be raised again with the addition of fresh warm water. Heat should not be applied directly to the tub since the frozen limb, insensitive to stimuli, might accidentally press against the heated part of the tub and get burned. The thawing process can take 30 to 60 minutes and, as mentioned, will be very painful. Pain medication should be given, if available.

After the limb has thawed, it must be elevated, kept clean to prevent infection, and protected from any form of trauma or irritation. When blisters form, leave them alone; they are a normal part of the healing process. The victim should be evacuated to a medical facility as quickly as possible.

Cold-water Immersion

If you fall into cold water, don't try to swim to safety unless shore or a boat is just a short distance away, or unless there is no possibility of rescue; water conducts heat from the body 250 times faster than air does, so you won't get far before being overcome by the cold. Instead, conserve heat by minimizing body motion. The Heat Escape Lessening Posture (HELP) consists of hugging your knees to your chest and floating with your head out of the water. This posture minimizes heat loss from major arteries in the groin, armpits, and neck. If you're

PLANNING FOR FIELD SAFETY

not wearing a life jacket—and you should be—scull gently with your arms to remain afloat.

When you first enter the water, you may find yourself unable to breathe normally; this is a common reaction to sudden immersion in cold water and will pass after a few moments of floating quietly in the HELP position.

The principal threat to life in cold water is drowning due to the paralyzing effects of hypothermia. An unclothed person will normally be helpless after less than one-half hour in water at 37°F and after about one and one-half hours in water at 47°F. With thick conventional clothing, the times before hypothermia affects the victim increase to approximately 40 to 60 minutes in water at 37°F and to about four hours in water at 47°F. The "5 rule" is an easy way to remember the dangers of cold water: a person has less than 50% chance of surviving five hours in 50°F water. Immersion time before hypothermia sets in can be significantly increased through using an immersion survival suit.

On boat trips, if boat gunwales are too high for someone in the water to climb easily into the boat, attach a short ladder to the side of the boat.

U.S. Coast Guard pamphlet CG 473, *A Pocket Guide to Cold Water Survival,* is recommended reading for those working on or around cold water.

Snow

See "Alpine Regions" for specific hazards associated with snowfields and glaciers.

Reflected ultraviolet rays can lead to severe sunburn in snow country, no matter what the temperature is. Be aware that on snow you can get sunburned in unusual places, such as under your chin and on the roof of your mouth.

Overexposure to the sun's ultraviolet rays can also cause superficial corneal burns, resulting in an extremely painful condition called snow blindness. This condition is marked by a feeling of sand in the eyes, excessive tearing, and swollen eyelids. The higher the altitude, the greater the exposure to these rays and the greater your risk of developing snow blindness. When working in snow, particularly at higher altitudes, wear dark sunglasses guaranteed to filter out ultraviolet radiation (the best of these tend to be expensive, but the eye protection they afford is well worth the price). Make sure your sunglasses have side shields to block out reflected light from snowfields. If you or someone in your group develops snow blindness, treat by bandaging the eyes and covering the bandages with cool compresses. Allow plenty of bedrest, and give pain medication, if available. Do not use any ointments or ophthalmic solutions, unless prescribed by a physician. Recovery is usually within 24 to 48 hours.

Hot Weather

When working in hot weather, drink frequently, before you feel thirsty. Monitor your urine output; if it is frequent and light colored, you are taking in adequate

PLANNING FOR FIELD SAFETY

amounts of fluid. In extremely hot weather or after great exertion, it's a good idea to drink Gatorade or some other glucose-based drink designed to replace minerals lost in sweat. Salt your food more heavily in hot weather to make sure you're replacing lost salt. Salt tablets, which can cause gastric irritation and vomiting, are generally not recommended.

Be careful not to overexert yourself. Take frequent rest and water breaks. In conditions of high humidity and high temperature, in which your body is unable to dissipate heat effectively, avoid strenuous activity altogether.

Use zinc oxide or lip balm containing a sunscreen to prevent chapped and sunburned lips. Don't forget to protect the backs and lobes of ears and the undersides of the nostrils—particularly if hiking on snow or sand.

Wear a broad-brimmed hat and sunglasses that filter out ultraviolet light to shield your eyes from the sun.

A good supply of clean clothes, especially socks, will help reduce your chances of skin irritation when sweating profusely. Boxer shorts cut down on crotch chafing. In hot weather, cotton breathes more readily than nylon and helps prevent heat rash. Prickly heat occurs when sweat glands become blocked. If possible, avoid creams that clog the pores.

Wear a light-colored, long-sleeved shirt and full-length pants in hot weather to reduce your risk of sunburn and heat illness. Moisture held in your clothing may feel uncomfortable, but it greatly reduces the rate of water loss in high-temperature environments. Long sleeves and pants will also make hiking through thick growths of sagebrush, cacti, manzanita, buckthorn, scrub oak, and other desert favorites a little less unpleasant.

Dehydration

Dehydration is a principal cause of heat illnesses. When the body loses more liquid than it takes in, it becomes dehydrated. Dehydration is not restricted to work in warm desert regions; it may also occur during fieldwork anywhere. Signs of dehydration include reduced and darker-colored urine, dry mouth, headache, and fatigue. To avoid dehydration, drink plenty of liquids before you set out in the morning, and continue to drink at frequent intervals throughout the day. If you're heading into the desert, it's a good idea to begin increasing your water intake a day or so beforehand. You should drink at least two quarts of water a day if you are active, more at high altitudes and under extremely hot or humid conditions. Remember to drink before you feel thirsty; thirst is not a reliable indicator of fluid status. Plan on needing at least two to three gallons of water per person per day for drinking, cooking, and cleaning, and have five to 10 extra gallons in camp at all times.

Cover plastic water bottles with heavy coarse cloth or burlap, or use cloth-covered canteens. When you wet the cloth, evaporation will help cool off the water inside. Desert water bags are useful for storing water at camp.

In the desert, drinking water requirements are several gallons per person per day. Keep in mind that one gallon of water weighs eight pounds. Your portable water con-

PLANNING FOR FIELD SAFETY

tainers must be able to hold enough water to last you over the distance you would have to walk for help. A couple of inexpensive, plastic one-gallon jugs of spring water, found in most grocery stores, is a minimum portable water supply.

Heat Illness

Normal human metabolism produces a tremendous amount of heat. Under normal operating conditions, the excess heat is carried by the blood to the surface of the skin, where with the help of perspiration, it is dissipated to the environment.

If for some reason your body is unable to dissipate heat effectively, usually because of overexertion in hot, humid weather, or a failure to replace lost water and salt, or both, you can develop hyperthermia, a condition marked by abnormally elevated body temperature. Hyperthermia results in two principal forms of heat illness: heat exhaustion and heat stroke.

Almost all cases of heat illness can be avoided by following the hot weather guidelines discussed, particularly maintaining adequate water and salt intake and limiting activity in hot and humid weather. Some people, such as the very young, the very old, and those with diabetes, cardiovascular disorders, and certain other illnesses, are especially susceptible to heat illness. Such people must be especially alert to the conditions that can lead to heat illness.

WEATHER CAUTIONS

Aspirin and related drugs should not be given in cases of heat illness. Not only are the drugs ineffective, they increase internal bleeding that commonly accompanies heat stroke.

Heat exhaustion. In an effort to dissipate heat in hot weather, the blood vessels of the skin may dilate to such an extent that the normal blood supply to the brain is reduced. Failure to replace lost fluids can compound the problem by decreasing the total volume of blood available. Such conditions, which can build up over the course of a day or over several days, can lead to a condition called heat exhaustion.

Symptoms. Heat exhaustion produces symptoms very much like fainting. The victim feels faint, has a rapid heart rate, and may experience a range of symptoms including nausea, vomiting, headache, dizziness, restlessness, and sometimes a brief loss of consciousness. He or she may be sweating profusely; skin may be ashen, and it may feel cool and clammy. Urinary output will be low.

Heat exhaustion is generally not associated with significantly elevated core body temperature. However, heat exhaustion can develop into heat stroke, a much more serious heat illness, so it is important to monitor body temperature closely.

Treatment. To treat heat exhaustion, lay the victim in the shade with feet slightly elevated. Give slightly salted water to drink (one-half to one teaspoon of salt per pint of water), about a glass hourly. If the victim's body temperature is elevated, continue to monitor it carefully.

PLANNING FOR FIELD SAFETY

If the temperature rises above 105°F, take immediate and aggressive steps to cool the victim according to the treatment instructions outlined for heat stroke.

After recovery, the victim should rest in the shade for at least a day. Only after body fluids have been completely replenished and salt and urinary output has returned to normal should the victim resume working in hot weather. Those whose body temperatures exceeded 104°F should probably be examined by a physician before resuming activity.

Heat stroke. When the body's normal cooling system—a combination of blood circulation and perspiration—breaks down in hot weather, a condition called heat stroke results. This disorder is far more serious than heat exhaustion. It can cause death or leave victims with permanent residual disabilities.

Symptoms. Heat stroke strikes rapidly. Its onset is marked by altered mental function, typically confusion and irrational behavior. Incoordination, delirium, and unconsciousness often follow, and convulsions are common. Pupils may be dilated and show no response to light. Any person with altered mental status or neurologic function in response to heat is considered to have heat stroke.

Other symptoms include profuse sweating, a rapid pulse, irritability, fatigue, and sometimes headache, dizziness, and nausea. As body temperature approaches 105°F, the victim will collapse. At this point, the victim's skin will probably be hot, flushed, and often dry, the sweating mechanism having broken down completely.

WEATHER CAUTIONS

Once this condition occurs, body temperature will rise to nearly 110°F in a matter of minutes, at which point the victim will die.

Treatment. Clearly, heat stroke is a serious medical emergency, requiring aggressive, immediate treatment. If a member of your party shows signs of heat stroke, get him or her out of the sun at once. Lay the victim down with feet slightly elevated. Remove clothing and fan the body to increase air circulation and evaporation. Cover the extremities and trunk with cool, wet clothes or towels, or pour cool water over the entire body. Continue to fan the body, and massage the extremities to accelerate the flow of cool blood back to the body core. In general, do whatever is necessary to lower the victim's temperature as quickly as possible.

When the victim's body temperature drops to about 102°F, scale back your cooling measures to avoid making the victim hypothermic. Continue to monitor temperature carefully; body temperature has been known to rebound suddenly in heat stroke victims three to four hours after cooling.

After recovery, the victim should be evacuated to a medical facility. There are many, often severe, complications of heat stroke, including kidney and liver failure, abnormal blood clotting, heart damage, and extensive brain damage.

PLANNING FOR FIELD SAFETY

Sunburn

Bad sunburn, particularly on your shoulders or legs, can incapacitate you. For many years, conventional wisdom held that the best way to avoid sunburn was to build up a base tan before heading into the field. But recent studies suggest that even moderate exposure to the sun can significantly increase your risk of skin cancer later in life. The only completely safe way to avoid sunburn, therefore, is to avoid exposure to the sun—by keeping to the shade as much as possible; limiting outdoor activities during the hottest hours of the day; wearing long-sleeved shirts, long pants, hats, and other protective clothing; and using a sunblock on exposed skin, with a protection rating of at least 15. Waterproof sunblocks are a good idea for those working on the water or in humid conditions. Note that the effects of the sun increase roughly 4% for every 1,000 feet of elevation, so particular vigilance against sunburn must be taken when working at high altitudes.

Thunderstorms

Lightning, flash floods, heavy rain and hail, strong winds, and tornadoes are the hazards posed by thunderstorms. If you're working in an area where thunderstorms are likely (nearly anywhere during the summer), keep an eye on the sky at all times. Watch for a sudden shift in wind direction or a sudden drop in temperature that might signal approaching bad weather. Learn to recognize the difference between mature, "anvil-topped" cumulonimbus clouds that signify a thunderstorm and puffy, fair-weather cumulus clouds. You can use the tails of flat anvil tops of cumulonimbus clouds (thunderheads) to tell which way a storm is moving. If you see a bolt of lightning, count the number of seconds until you hear the thunder. Divide this number by five to determine roughly how many miles away the storm is. Repeat this procedure to figure out if the storm is moving your way, and if so, how quickly.

If a storm is on the way, seek shelter immediately. Get away from areas such as exposed mountaintops, ridges, and plains. If you're in a low-lying area or canyon where the threat of flash flood exists, get to higher ground at once.

In mountainous areas, heavy rains at higher elevations may generate flash floods in the lower canyons. If flooding is suspected, do not attempt to drive your vehicle out of the canyon if you are any distance from the canyon mouth. Abandon your vehicle and seek shelter up the canyon wall, but do not use a side canyon as a shelter.

Lightning

By far the most dangerous product of thunderstorms is lightning. But in truth, lightning poses very little threat to those who keep their basic earth science in mind, keep an eye on the weather, and exercise a little common sense.

PLANNING FOR FIELD SAFETY

For those who need a quick review, lightning is simply the electrical arc that forms when current jumps from the positively charged earth to a negatively charged cloud—very much the way a spark jumps the gap in a spark plug. It stands to reason, therefore, that the closer the two contact points—earth and cloud—are to each other in a particular area, the more likely an electrical discharge will pass between them. Thus, objects rising up higher than the surrounding landscape are very often the first things that lightning will strike.

If a lightning storm catches you in the open, take immediate steps to make yourself a poor lightning rod. Get off mountaintops and below the tree line as quickly as possible. If you're caught in an open field, head for the surrounding woods, if possible, or crouch down in a depression until the storm passes over. Don't lie down; the idea is to minimize contact with the ground. Never stand under a prominent object, such as a tree in the middle of a field; stay away a distance equal to about twice the height of the object. Stay clear of metal objects, such as pack frames, tent poles, metal stadia rods, and soil augers. Don't handle open containers of flammable liquids. If you take shelter in a cave or under an overhanging rock, stay as far back from the cave's entrance or the outer edge of the rock as possible; lightning often jumps gaps of this sort and will pass through anyone standing in the way. If you're on a lake, get to shore as quickly as you can. Never swim during a lightning storm. If you're in camp when a lightning storm

WEATHER CAUTIONS

approaches, disconnect radio antennas until the storm passes.

In the rare event that someone in your group is struck by lightning, perform CPR if necessary (continue CPR as long as you are physically able; victims of severe electrocution have been known to show no response for several hours and then make a full recovery), and treat all wounds as burns.

Lightning is also the principal cause of forest, grass, and other wildfires—themselves serious threats to fieldworkers.

6

ANIMALS AND PLANTS

Animals

Bears

For fieldworkers stationed in bear country, the threat of bear attack cannot be taken lightly. However, by learning some basic facts about bears and their habits, and by taking sensible precautions to minimize bear encounters, fieldworkers can reduce that threat considerably.

Range and habits. Black and grizzly (brown) bears range from the ocean beaches to alpine ridges. Black bears (*Ursus americanus*) live in or very near forests and avoid open spaces. Grizzlies (*Ursus arctos,* formerly *Ursus horribilis*) prefer open country, but do not hesitate to travel through and to forage in the forests. The polar bear (*Ursus maritimus*) is usually found on the Arctic Sea ice, but it dens and sometimes seeks plant food and carrion on land.

Polar bears are primarily meat eaters. Black bears and grizzlies eat mostly plant foods—berries; nuts; and herbaceous plants, roots, and tubers—but supplement their diet with fish, carrion, and fresh game (usually rodents). Therefore, places you're most likely to encounter black or grizzly bears include salmon-spawning streams, lush meadows, river bottoms, avalanche chutes, old burns, berry patches, and areas with numerous rodents, such as highway edges, power line right-of-ways, and alpine ridges.

Possessing a keen sense of smell, bears can scent decaying carcasses miles upwind, and they often patrol tidewater beaches for tidbits. All are possessive, often ferociously so, of carcasses and kills and even of gut piles left by hunters. Bears commonly cover or partially bury such prizes for later consumption. There is evidence that a few bears of all species, most commonly subadult, and perhaps subadult males more commonly than subadult females, attack humans as prey. Black bears and grizzlies enter dens in late October or November, where they normally remain until mid-April or May. They do not truly hibernate, but are in a dormant state, and if disturbed can react dangerously almost immediately. Bears found moving about during these months may be sick or hungry and so may be especially dangerous.

The usual response of bears to potential close encounters with humans is to avoid them or run away. However, bears are naturally curious and are unpredictable, even to specialists in bear behavior. Some bears will attack if they are surprised, if their personal space or tolerance range is encroached upon, or if they feel that their cubs or food is being threatened. While bears will not ordinarily leave their cubs to attack strangers, intruders who get between a sow and her cubs will commonly be charged. Bears have as many as four cubs a year, but two is the average. Cubs stay with their mother from birth in January or February to 15 or 16 months later, at which time she usually drives them away and mates again. Black bears are good climbers; they will sometimes take refuge in trees, especially if accompanied by cubs. Among grizzlies, tree climbing is usually done only by subadults, but adults can climb trees with widely spaced strong limbs. Partly because they are larger and partly because they do not usually have the tree-climbing option, grizzlies and polar bears are more likely to attack people than black bears are. However, black bears are more plentiful, and they do sometimes attack humans.

The usual avoidance behavior of bears toward humans is altered in places where contact with humans is frequent, such as in state and national parks. In such situations, some bears learn to associate humans and their odors with easily obtained food. The association occurs at camps, cabins, caches, garbage dumps, and anywhere that tourists purposely feed bears. Such human-conditioned bears are especially dangerous because they develop the notion that a human is easy prey, and they sometimes act on that notion. As a consequence, bears near settlements are probably more dangerous than those in remote areas.

Avoiding bears. Minimize the chances of sudden close encounters with bears by being alert and aware of your surroundings. Make a habit of scanning ahead and behind for bears. Be particularly watchful upwind, as bears, unable to sense your approach, are more likely to be present there. Bears feel threatened by surprise. Where possible, choose routes across open areas with good visibility. Avoid patches of willow, dwarf birch, tall grass, and other likely bear daybed areas. Watch for fresh signs of bears, such as tracks, droppings, holes dug in search of rodents, uprooted logs, and clawed or rubbed tree bark. Be wary of strange smells; bears give off a strong odor, especially when excited. Avoid or be especially alert near berry patches, salmon streams, and other areas with abundant bear food. Never tarry near carrion, fresh kills, or gut piles, whether on the surface or partially buried. Remember that bears, like people, prefer to take the path of least resistance, so they will usually choose the same routes and trails as you across rough terrain. If you have alternatives and bears are numerous, avoid main trails, beaches, and stream banks.

Where visibility is impaired, as in brush or woods, and especially when walking upwind or crosswind, try to give bears advance warning of your approach by making noise: talking, wearing bells, snapping sticks, banging rocks, or rattling pebbles in tin cans. When traveling in brush, strong wind, or near rushing water, however, don't count on being heard.

If you come across a bear cub, do not approach it, even if it appears to be alone. Walk back the way you came, and continue on, using an alternate route.

ANIMALS AND PLANTS

When bears are in the vicinity, and especially when working in tundra or places of limited visibility, consider carrying a suitable firearm. A .30 to .06 caliber or larger rifle loaded with 180 or 220 grain soft nose bullets or a 12-gauge shotgun loaded with slugs is suitable. Hand guns are not recommended because they lack the necessary killing power and accuracy except at point-blank range; they may provide a psychological crutch that encourages the bearer to handle encounters more casually than otherwise. If using a firearm, be properly licensed and trained to do so. Learn and scrupulously follow accepted firearm safety guidelines, and follow all local firearm restrictions.

Remember that using firearms in defense of life or property is an absolute last resort. Be aware of the risk to other party members and of the dangers posed by a wounded animal.

It is illegal to kill protected species except in life-threatening emergencies. It is also illegal to kill black bears without a license or out of season. If you have to shoot a bear, report the incident to the fish and game department in your area. If you kill a grizzly in some jurisdictions, you must bring in the hide, skull, claws, and gallbladder when you report the kill. These items are required as evidence that you did not kill the bear for sport or that you were not poaching.

Avoid carrying or using scented creams and lotions, such as cosmetics, deodorants, hairspray, insect repellents, and suntan lotions. These may attract, or even infuriate, bears. Hang these items out of bears' reach at

PLANNING FOR FIELD SAFETY

night or when away from camp. Where hanging is not an option, store these items well away from camp.

There is some evidence, although not conclusive, that persons who have recently had sexual intercourse or who are menstruating are more likely to attract bears. Scrupulous personal hygiene and extra caution are advisable at these times.

Avoid carrying smelly foods. Food in tins, sealed foil, or heavy plastic containers is least likely to invite a bear's attention. Keep food in your pack, not in your pockets.

Always carry your medical kit.

Working in pairs rather than alone probably decreases the chances of a bear attack, although a group of four or more is a more reliable deterrent.

If you're working with helicopters or other close logistic support and you are within range of a radio base station or relay station, carry a radio where you can get to it easily in an emergency. Don't hesitate to use it early in a bear encounter, before the situation has a chance to get out of hand. The combination of a monitored radio station and field support helicopters has saved the lives of at least two people severely mauled by bears. If a monitored radio base is not available, establish a radio schedule with your helicopter, radio base, or other members of the field party. Keep in mind that bears can easily be scared off with a helicopter (but don't harass bears with a helicopter unnecessarily; bears so bothered have been known to destroy grounded helicopters at night). If possible, fly over your route to check ahead for bears before you begin your traverse. Be on the lookout for bears before landing, and have firearms and ammunition readily available when you're dropped off. Setting down along a stream might put you right in a bear's front yard.

Camping defensively. Don't camp on game trails, near good bear feed, such as berry patches or salmon streams, or where there are fresh signs of bears. Beaches, mountain passes, ridge crests, and river banks are favored bear paths. Bears often travel at night, and many bear incidents at camp occur during the early morning hours.

When you leave for the day, don't leave firearms in camp. A bear might visit camp while you're away, and you may be unable to get back to your weapon. Consider storing it in a waterproof gun bag in a location that is on your way back into camp.

In wooded country, particularly if you don't take along a firearm, camp where visibility is good and near a large tree that you can climb if necessary.

Set up the cook tent or other food preparation area well downwind of the sleeping area—the farther the better—and don't let anyone sleep in the cook tent. Any clothing that has been soiled with food, game, or fish should be kept out of sleeping tents. Never eat in your sleeping tent or sleeping bag.

Use relatively odor-free and nongreasy foods, such as freeze-dried or dehydrated food, and keep all food in sealed containers. Store foods; other aromatic items, such as lotions, toothpaste, shampoo, and cosmetics; and clothing soiled with food in bear-proof caches or metal

PLANNING FOR FIELD SAFETY

boxes, or suspend them in bags on a rope stretched between two trees at least 12 feet above the ground. If you have to store unprotected food on the ground, put the food well downwind of your sleeping area.

Burn all garbage, especially food waste and packages. Do not bury the noncombustible residue; store it in the same manner as you store your food, and pack it out. Good practice is to burn the garbage several hundred yards downwind from camp. Burn garbage only in areas where forest fire or tundra fire danger is minimal. Otherwise, carry all garbage out.

Don't leave dirty pots or eating utensils around camp, and do not clean fish or game near camp.

Never feed bears, or leave food out for bears, birds, or any other wild creatures. If there is a dog in camp, treat the dog's uneaten food like garbage—do not leave it around overnight.

Sleep in a tent and try not to overcrowd it. There have been quite a few incidents where bears have attacked people who were in sleeping bags in the open or who were lying against the walls of crowded tents. Keep a knife handy in case you need to cut a quick exit hole.

Use a flashlight to look around carefully before leaving the tent at night.

If you meet a bear. If you see a bear before it sees you, stop and plan a route around it. Give the bear all the room you can, and do not trespass on its personal space, which can extend outward for several hundred yards. If possible, keep the bear downwind to let it know you're there. Bears reportedly prefer to retreat uphill, so try also

ANIMALS AND PLANTS

to stay downhill of bears. Bears have good senses of smell and hearing, but reportedly have difficulty discriminating stationary objects beyond about 300 yards.

If you meet a bear at close range, stop. Above all, don't panic or run; the bear will instinctively chase you and will easily catch you. Don't shout or make threatening or aggressive sounds or gestures. Don't make abrupt movements. Take comfort from the well-established fact that in almost all cases the bear will back down or run away. The signs a bear makes when it feels threatened—walking about stiff-legged, pricking up its ears, growling, urinating, stamping the ground—are all indications that the bear has not yet decided to attack. If it plans to attack, it will simply do so, coming toward you at a fast walk or run. However, some charges are bluffs, terminating a few feet short of contact. It is this unpredictability that makes dealing with bears so difficult, and so dangerous.

If the bear is close enough to hear you, talking to it in a soft, monotonous voice is supposed to help communicate the idea that you have no aggressive intent, thus lessening the chances that it will attack in self defense. As you talk, back slowly away from the bear.

If the bear follows, drop pieces of clothing and equipment one at a time as you retreat; with luck, these will divert the bear's attention. Unless it contains food, keep your pack on; it may provide some protection should the bear actually attack.

If the bear continues to pursue you, there's no sense continuing your retreat. Stop. Draw yourself up into as

PLANNING FOR FIELD SAFETY

imposing a figure as possible, wave your arms, and shout as loudly as you can. The idea is to convince what may be simply a curious bear that you are not its usual prey, in the hope that it will be discouraged.

If you are unarmed and it is clear the bear is going to attack, drop to your knees, press your face against your thighs, and clasp your hands over the back of your neck. This position will help protect vulnerable body parts. Stay as still as possible. In many such cases, bears have run right by a person or inflicted only minor injuries. Most mauling victims survive.

Playing dead, however, is probably not a good idea if the bear appears to have aggressive intent (e.g., it appears to be sick or wounded or has been stalking or following you). Continue to shout at the bear, and try to strike it in the face, especially on the top of the nose, with a hammer, ax, club, or whatever is at hand, hoping to hurt and frighten or outbluff it. These actions have reportedly been done a number of times, sometimes with great success . . . and sometimes not.

If you have a firearm, don't use it until you are sure you have no other alternative. A wounded, adrenalin-charged bear is so dangerous that shooting it should be your absolute last resort. Wait until the attacking bear is within 20 yards, preferably within 15 yards, before squeezing the trigger because the bear may be bluffing. If it is not bluffing, the close range will increase the chance that your first shot will be disabling. The first shot has to be your best. Be certain that no one is in the line of fire. Bears attack on all fours, walking or running in a low crouch. They don't charge on just their hind legs. They can charge as fast as 10 yards per second. When they rear up on their hind legs they are trying to smell, hear, or see you better, and this maneuver is at least as often followed by retreat as by a charge. Aim for just below the snout in the chest region. Note particularly that the front and top of the head are not suitable targets when a bear is coming at you, which is presumably almost the only time you will shoot one. If you wound a bear, you have the responsibility to try to kill it. Don't try to kill the bear alone; get help. Be extremely careful, particularly if you must follow the bear into brush where it can squat and be very hard to see until you are only yards away. Wounded or frightened bears almost invariably head for alder or willow brush where they feel most at home, where their superior senses of hearing and smell are more useful than your superior eyesight, and where you will usually have difficulty walking and handling a firearm. No one without a suitable firearm should help track down a wounded bear.

Dogs as bear guards. A trained dog can be an excellent early warning system for bears, particularly at night when you may be unaware that a bear is in the area. A dog's barks will generally keep a bear out of camp. If a dog is sufficiently large and aggressive, it can also deflect a bear attack; lead or chase a bear away from you or your camp; or distract a bear long enough to permit you to get out a gun, climb a tree, or otherwise escape. On the other hand, untrained dogs may attract or infuriate bears, precipitate attacks, and lead angry bears right to you or your camp. If you are sure that your dog will behave, it can be a valuable addition to your field party, but weigh the risks before you bring it along.

Other Animals

In northern regions, watch out for bull moose during the rut, mid-September through October, and for cow moose with calves; they are easily provoked and should be avoided. Elk and deer can also be dangerous in these situations.

Unfriendly dogs can be a hazard anywhere. Keep an eye out for them, particularly if you're crossing private land. Wolves, coyotes, foxes, skunks, raccoons, and other small animals are usually afraid of humans, but they may attack under certain circumstances, particularly if they feel cornered. Don't try to approach them, even if they act tame; when a wild animal appears unusually friendly, there's a good chance it has rabies or some other illness that is affecting its mental stability. Occasionally, squirrels have been known to bite humans; however, they have not been found to carry rabies.

In some cliff areas a hard hat may be useful to ward off attacks by eagles and hawks defending their nests, or by arctic terns at high latitudes.

In alligator or crocodile country, don't hike at night; these animals often travel on land after dark. Obviously, don't swim in waters infested with these creatures.

Cougars rarely attack humans unless they are defending their young or are old and starving. If you encounter

ANIMALS AND PLANTS

one of these animals, don't turn and run or attempt to climb a tree; it will instinctively chase—and catch—you. Instead, face the animal, stand tall, and back away slowly. If it pursues you, swing at the animal with any available object in an attempt to inflict pain; cougars are said to have a low tolerance for pain and will probably break off an attack rather than come back for more.

Snakes

Contrary to popular belief, snakes do not spend their days lying in wait for unsuspecting fieldworkers. They are, in general, timid creatures who avoid contact with humans whenever possible, normally striking only when startled or frightened. The vast majority of snakes are not venomous. Of the many types of snakes found in the United States, for example, only four carry venom: the coral snakes (found in the Southeast and Arizona) and the three families of North American pit vipers (named for the small pit between eye and nostril), which include the copperhead (found throughout the East), rattlesnakes (widespread), and the water moccasin or cottonmouth (found in wetlands of the South).

This is not to say, however, that the threat from snakebite is not a serious one; it is. In some parts of the world, snakes exist whose bite can be fatal within minutes. All fieldworkers heading into snake country (particularly the tropics, where venomous snakes are more common—and more deadly) should learn to identify and

PLANNING FOR FIELD SAFETY

avoid the types of snakes they are likely to encounter. Avoiding snakebites is far easier than treating them.

Habits. Although each species of snake has its own peculiarities, all are governed by a few physiological rules that can help you determine when and where you're most likely to find them.

Being coldblooded creatures, snakes must regulate their exposure to the environment to maintain a stable internal body temperature. On cool days you are most likely to find snakes sunning themselves on stretches of open trail, and on cool nights, soaking up the trail's residual warmth. In warmer weather, snakes retreat to the cool shade of bushes, rock crevices, tall grass, and other sheltered places.

Always keep an eye out for snakes. Watch especially where you put your hands and feet. Don't step over fallen logs without first checking for snakes on the other side. When climbing slopes or collecting rock samples, don't put your hands or feet into any crevices or hollows or onto any ledges until you are sure there is no snake lurking there (and don't hold your face too close while checking). Don't turn over rocks with your bare hands; use your rock hammer, and roll the rock toward you—putting the rock between you and whatever may be coiled beneath. Travel as little as possible after dark, when snakes are most active. Don't collect firewood after dark. Wear long, loose-fitting pants and sturdy leather boots that cover your ankles. Small coral snakes have been known to curl up in warm boots at night, so it's a good idea to keep your boots inside your tent. Never try to capture or kill a snake, unless you need a record of the culprit for medical purposes.

Although technically deaf, snakes are highly sensitive to vibration. When traveling in snake country, announce your approach by talking loudly and treading heavily.

If working in snake country during winter months, be aware that snakes hibernate, sometimes in dens of up to 100 or more snakes. Before heading out, check with local authorities on the locations of these dens to avoid discovering them on your own.

If you come across a snake. Stop. Make no sudden motions that might cause it to strike. Back slowly away until you're safely out of the snake's range (and this doesn't mean all the way back to civilization; even when tightly coiled, a snake can strike no farther than the length of its body). Give the snake a wide berth when you continue on, and watch for other snakes in the area.

Treating snakebite. If you or someone in your party is bitten by a snake, don't panic. Chances are the snake wasn't poisonous or, if it was, didn't inject venom. Have the victim relax in the shade until a course of action can be determined.

Your first step should be to identify the type of snake involved, whether by examining the snake or, if this is not possible, by examining the puncture marks it left behind. Fieldworkers in snake country must be well versed in local snake identification. Proper identification will enable you to determine immediately whether or not the snake was poisonous, and thus whether a health threat exists. In some foreign countries a separate antivenin is used for each species of snake, so proper identification is essential for proper treatment. In the United States, a single antivenin is used to treat all pit viper bites, and another antivenin is used for all coral snake bites.

If the snake was poisonous, you must next determine whether or not it injected venom. A fieldworker in the United States is most likely to be bitten by a pit viper; coral snakes are shy and rarely attack humans. Pit viper envenomation will become obvious almost immediately, with pain, local swelling, and perhaps a metallic or rubbery taste or tingling sensation in the mouth. If such symptoms do not occur within 15 minutes of a pit viper bite, it is generally safe to assume that envenomation did not occur. With many other species of poisonous snakes, however, including coral snakes, signs of envenomation may not become apparent for up to six hours after the bite. Thus, if you determine the bite was from a coral snake or another venomous snake whose effects are not immediately apparent, or if you are unable to make a positive identification, assume envenomation took place.

Regardless of the type of snake involved, if envenomation has occurred (or may have occurred), immediately evacuate the victim to a medical facility for antivenin treatment. Ideally, the victim should be transported on a litter, as any movement on his or her part will hasten the spread of poison throughout the body. If the victim must be walked out, do so slowly, keeping the victim cool and taking frequent rest stops. Upper extremities may be splinted if doing so will make the victim more comfortable.

Treatment of snakebite in the field is not recommended. "Cut and suck" snakebite kits, once commonly used, have proven ineffective at best, and dangerous at worst. Likewise, the use of tourniquets or pressure wraps to constrict lymphatic flow and reduce the spread of venom has not been shown to have any significant value and may result in increased local tissue swelling and damage. Carrying antivenin into the field is also not recommended. Antivenins are made from horse serum, and in those sensitive to horse serum (up to 70% of the population), reaction to the antivenin may be more life-threatening than reaction to the venom itself. Antivenin administration is quite difficult, requiring a series of subcutaneous injections. In the case of pit viper bites, for example, up to 40 vials of antivenin may be required—far more than any field party would have on hand. For these reasons, antivenin injections should be given by a physician only, in a hospital setting.

Never give a snakebite victim alcohol in an attempt to calm him or her down; alcohol accelerates venom absorption. Never apply ice to a snake wound in an attempt to reduce swelling. Such action causes venom to concentrate in one spot, increasing the risk of severe tissue damage. Be aware that snake venom is often more concentrated in the spring, when snakes are fresh from hibernation, and can cause more severe reactions than at other times of the year.

Even if a snakebite victim was not envenomated, a physician should be seen to prevent infection.

Insects and Other Pests

Mosquitoes and Flies

Unless you confine your fieldwork to cold weather seasons, you will contend with your share of mosquitoes, flies, no-see-ums, and other annoying insects. Carry an insect repellent with you. Be aware that some repellents may damage some plastics and synthetic fabrics, such as the mosquito netting on your tent. In some areas where these insects are particularly bad, wear a headnet. Anyone who's ever been driven to the brink of madness by swarms of deerflies and the like knows how seriously these pests can impair judgment and limit safety awareness.

Some people are mildly allergic to black fly or mosquito bites and may become progressively more allergic with each exposure. A mild antihistamine should be carried in such cases.

Mosquito-borne diseases. In some parts of the world, particularly in Central America, South America, the Caribbean, sub-Saharan Africa, Southeast Asia, and the Indian Subcontinent, mosquitoes can transmit several potentially fatal diseases. The most common of these is malaria, whose flu-like symptoms include fever, chills, headache, and general malaise. One type of

ANIMALS AND PLANTS

malarial infection can cause kidney failure, coma, and even death.

When working in areas where malaria might be a problem, carry chloroquine tablets (consult your physician for the necessary prescription). Start taking the pills two weeks before your trip and continue for six weeks after. Be aware, however, that using these pills does not guarantee safety from malaria. Contact the Centers for Disease Control at 404/639-3534 for information on regions of chloroquine-resistant malaria.

Try to limit your exposure to mosquitoes through the use of repellent and mosquito netting, particularly at night when anopheles mosquitoes, the principal carriers of malaria, do most of their feeding. Watch for signs of malaria (which may not appear until up to eight days after infection) in yourself and others. If you suspect you may have contracted the disease, see a physician. Malaria is much easier to treat if caught in its early stages.

Another common disease transmitted by mosquitoes is dengue (or breakbone) fever. Although not a fatal illness, dengue fever can be very painful. High, long-lasting fever followed by an overall rash is typical. Time is the only cure for dengue fever, which may take up to 10 days to run its course. See a physician if possible.

Some types of encephalitis, a sometimes fatal inflammation of the brain marked by headache, neck stiffness, vomiting, and diarrhea, can be transmitted by mosquitoes. Japanese encephalitis may occur in epidemics throughout parts of Asia in late summer and autumn. Other types are occasionally found in Canada, Alaska,

PLANNING FOR FIELD SAFETY

and the contiguous United States. If you plan to be working for an extended period—a month or more—in an area where exposure to this disease is possible, contact the Centers for Disease Control for information on vaccination.

Bees, Wasps, and Ants

Painful but not terribly serious for most people, the stings of bees, wasps, and certain types of ants can cause a potentially fatal reaction, called anaphylaxis, in those highly sensitive to their venom. Such a reaction is marked by hives, labored breathing, weakness, and other symptoms that can result in death in a matter of minutes in particularly severe cases. Those subject to such reactions should carry a bee sting kit (as it's generically called) whenever they travel outdoors in warm weather. These kits, available by prescription, contain a syringe of epinephrine, which reverses severe reactions, and an antihistamine, which helps prevent the reaction from rebounding when the effect of the epinephrine wears off. All members of the group, whether allergic or not, should be familiar with the contents of the bee sting kit and know how to use it. Premeasured injectable epinephrine "pens" are also available and may be easier for some people to use.

If you're not sure whether or not you're allergic to bee stings, have a physician test you. Be aware that people become progressively more allergic to bee stings over the years, so if you've experienced a fairly severe reaction

ANIMALS AND PLANTS

in the past, chances are your reactions will only get more severe with age.

Observant fieldworkers should have little trouble with bees and wasps. Bees tend not to sting unless provoked. Be careful not to step or sit on bees or trap one in your clothing. Don't walk barefoot in areas likely to attract bees, such as around trash cans or through fields of flowers. When a bee stings, it will often leave its stinger behind. Be sure to scrape the stinger out with a knife blade rather than pull it out with your fingers; this tends to squeeze more poison into the wound.

Wasps, especially hornets and yellow jackets, will often attack if you venture too near their nests. Keep an eye out for gray paper nests hanging from trees and for wasps hovering near the ground or around rotted stumps or food. Give these areas wide berth. Food and beverages attract yellow jackets, so don't leave garbage out.

If you are stung by a bee or wasp, cool soaks will help relieve some of the discomfort, and Adolph's Meat Tenderizer worked into the wound is often effective at reducing the pain and swelling. However, if a sting victim begins to show signs of an anaphylactic reaction, use the bee sting kit immediately—before the reaction progresses to a critical condition.

"Killer bees" are not as deadly as the term implies. For more information on them, see the article by Sue Hubbell in *Smithsonian*, listed in "Information Sources."

PLANNING FOR FIELD SAFETY

Spiders

Three types of spiders, found throughout temperate regions, pose a potential threat to humans: tarantulas, black widows, and brown recluses. Fieldworkers should learn to identify these spiders and keep an eye out for them when moving rocks or other objects and taking soil samples. Don't invite trouble by sticking your hands into dark crevices or holes. Shake out boots and other garments that were left outside overnight.

The bite of a tarantula is normally no more serious than a bee sting. Since tarantulas seldom venture out of their dens until late fall, few careful fieldworkers are likely ever to see one, let alone be bitten.

Black widows are more of a problem. Their bite often feels like little more than a pinprick, but within three hours can cause severe pain and muscle cramps that last up to four days. There is no effective field treatment for a black widow bite, other than keeping the victim cool and relaxed to slow the spread of venom. Fortunately, virtually all healthy adults will survive the bite without treatment, but victims should nonetheless be evacuated to a medical facility as soon as possible, where antivenin treatment is available.

The bite of a brown recluse, although practically unnoticeable at first, slowly develops into a gangrenous sore that may slough off skin for more than a year. The bite may also cause fever, vomiting, rash, and joint pain and may be fatal to children. There's no effective treatment for a brown recluse bite short of cutting out the infected area. This operation should be performed by a physician only, so get the victim to professional help immediately.

Scorpions

Members of the spider family, scorpions are found throughout hot, dry climates. The sting of most species, contrary to desert legend, is seldom more serious than a bee sting, causing intense pain and local swelling, but not death. Only one species of scorpion, *Centruroides sculpturantus,* a small, sand-colored variety found in New Mexico, Arizona, along the Colorado River in California, and throughout Mexico, is potentially lethal. Victims of this scorpion should be evacuated to a medical facility immediately.

With care, however, and some knowledge of their habits, fieldworkers should be able to avoid scorpions. During the day scorpions seek shelter from the sun in the sand, under rocks, or under dry leaves. Be sure to turn over rocks with your rock hammer, and don't stick your hand into any crevices. Watch where you sit if there are a lot of dry leaves or other litter around.

Scorpions tend to forage at night, so keep your tent zipped shut and check inside boots and clothing in the morning for unwanted guests. Also, check in open cups and glasses before filling them.

Ticks

Although most ticks are no more than a disgusting nuisance, some can transmit Rocky Mountain Spotted Fever (a misnomer, since 95% of cases occur east of the Mississippi), a potentially fatal illness marked by persistent fever, headaches, chills, rash, and sometimes death; and Lyme disease, which can cause crippling arthritis, usually of a single joint. If diagnosed early, both of these diseases can be effectively treated with antibiotics, but the best treatment is to find and remove ticks before they have a chance to transmit disease.

Ticks are usually found in fields of tall grass or similar vegetation, waiting patiently for a warm meal to pass by. Once on your body, they crawl to soft, warm areas, burrow in with their heads, and begin sucking blood.

When in tick country, make access to your skin as difficult as possible: wear long pants and a long-sleeved shirt, preferably of some light color to make spotting attached ticks easier. Apply an insect repellent to boots, the tops of socks, the cuffs of sleeves and pants, and other areas to steer ticks away from potential points of access.

Check your body and clothing frequently for attached ticks (group members should also check each other). Careful checking is especially important for those at the front of a hiking party. The longer a tick is attached to your body, the more difficult it is to remove and, if the tick is infected, the more likely it is to transmit disease. Each night make a thorough "tick check" of your clothing and body, especially the groin area, the base of the neck, and behind the ears. If you find one tick, carefully recheck for others.

If you find a tick, gently remove it with tweezers by pulling it straight back and off the skin. Be careful not to leave the head embedded, since the head can still transmit disease. If the head does remain behind, it should be cut out by a physician as soon as possible. Other methods of tick removal are popular—burning the tick with a cigarette or covering it with oil—but most physicians recommend gently removing it with tweezers.

If a member of your field party develops an unexplained fever when working in tick country, have him or her tested by a physician for Rocky Mountain Spotted Fever and Lyme disease as soon as possible.

Chiggers

These tiny mites, while not a serious danger, burrow under the skin and cause severe itching and discomfort. They generally form a pattern along the elastic portion of underwear, although they can also be found on legs and arms. The best treatment is a dab of nail polish on the affected area to suffocate the chiggers and prevent their spread.

Poisonous Plants

Learn to recognize common poisonous plants—for example, poison ivy, poison sumac, poison oak, stinging

PLANNING FOR FIELD SAFETY

nettles—and avoid them. If you accidentally touch one of these plants, immediately wash the affected area with soap and water. If you're particularly allergic to these plants, be sure to carry cortisone cream, calamine lotion, or some other treatment with you. A very severe reaction can be incapacitating and may require professional medical treatment with systemic steroids.

Stalking wild asparagus is not recommended. Many common plants are poisonous if ingested in large enough quantities, so unless you're absolutely sure that a plant or mushroom is safe to eat, stick to your trail food.

7
HAZARDS OF SPECIFIC REGIONS AND SETTINGS

This chapter provides guidelines for safe fieldwork in specific work settings and geographical areas. Because it focuses primarily on the safety considerations unique to these settings, fieldworkers should be sure to read the general sections elsewhere in this guide that pertain to their trip.

High Altitude

When climbing or working at elevations above about 8,000 feet, decreased atmospheric pressure reduces the amount of oxygen carried in the blood. At an altitude of 16,000 feet, this amount is reduced to one-half of that normally carried. Such a reduction can lead to vascular changes in the brain and lung, resulting in a range of uncomfortable or dangerous conditions broadly classified as altitude sickness. The relatively common altitude sickness, acute mountain sickness (AMS), has unpleasant symptoms in its mild and moderate forms. The severe forms of AMS are high altitude pulmonary edema

(HAPE) and high altitude cerebral edema (HACE); both can be fatal.

The effects of altitude on the body are significantly increased at higher latitudes because of decreased atmospheric density. The effects caused at a certain elevation on Denali (Mt. McKinley), which is about 63°N, for example, would be about 15% greater than the effects at the same elevation on Sagarmatha (Mt. Everest), which is about 28°N.

Acclimatization

The worst symptoms of altitude sickness can be mitigated or avoided altogether by giving your body time to adapt, or acclimatize, to the new environment. Acclimatization to high altitude produces several physiological changes: increased depth and rate of respiration, which increases respiratory volume; increased pulmonary artery pressure, which maximizes the ability of blood vessels in the lungs to absorb oxygen; slight increase in the number of red blood cells (which has not yet been proven to be significant); an increase in a blood enzyme that facilitates the release of oxygen from hemoglobin to body tissues; and changes in body tissues that allow them to operate at very low oxygen pressures.

When acclimatizing yourself to high altitude, there is no substitute for time; vitamins, medications, even previous aerobic conditioning will not speed up the adaptation process. The time required for the physiological changes to take place varies for each person, but approx-

HAZARDS OF SPECIFIC REGIONS AND SETTINGS

imately 80% of the adaptations occur in six to eight days, 95% in about six weeks.

The best way to acclimatize yourself is to ascend to high altitude gradually, stopping and resting for several days at intermediate elevations. The number of intermediate stops you make and the amount of time you spend at each will vary depending on the altitude of your final destination. If you're planning to work at 10,000 to 14,000 feet, for example, you might climb first to an elevation of 6,000 to 8,000 feet and spend three to four days there, exercising mildly to begin the acclimatization process. You would then continue on to your final elevation, resting there a few days before beginning any strenuous work.

At altitudes above 14,000 feet, you should limit daily climbs to 1,000 feet a day, resting every third day. It's also a good idea to drop down a few hundred feet at night to camp. There is evidence that a lower elevation for sleeping reduces the effects of sleep hypoxia, a condition caused by a decrease in arterial oxygen during sleep; at high altitude, these effects can make sleep difficult and aggravate the symptoms of acute mountain sickness.

Remember to drink plenty of fluids—several quarts a day—dehydration is a constant concern at high altitude. Insufficient food intake is another common problem, perhaps due to loss of appetite. Fieldworkers at high altitude should make a conscious effort to maintain an adequate diet high in carbohydrates.

Acute Mountain Sickness

Acute mountain sickness (AMS) usually occurs only at elevations above 8,000 feet, although some people have developed the condition at lower altitudes. The exact symptoms and their duration vary, depending on the elevation, the degree to which an individual has acclimatized himself or herself, and individual susceptibility.

Mild to moderate symptoms commonly include headache, nausea, dizziness, fatigue, shortness of breath, loss of appetite, and trouble sleeping. Generally, these symptoms appear 12 to 36 hours after reaching a new elevation.

Those suffering from acute mountain sickness should avoid strenuous exertion, but should continue to exercise lightly to promote acclimatization. Sleep, which slows respiration rates, tends to exacerbate rather than help the condition. Victims should drink plenty of liquids, eat foods high in carbohydrates, take mild pain relievers for headaches, and avoid alcohol, tobacco, and sedatives. Immediate descent of 1,000 to 2,000 feet is recommended for moderate AMS, 2,000 to 4,000 feet for severe AMS.

The drug acetazolamide (Diamox) has been shown to ease the symptoms of acute mountain sickness. Keep in mind that it needs to be taken beginning several days before a climb. Acetazolamide is available by prescription and should be taken only under the guidance of a physician.

HAZARDS OF SPECIFIC REGIONS AND SETTINGS

Symptoms of Severe AMS

Each form of severe AMS, high altitude pulmonary edema (HAPE) and high altitude cerebral edema (HACE), can be fatal. Thus, recognizing the symptoms early and getting medical help are vital. Symptoms of moderate AMS, such as severe headache and fatigue, may become more extreme in severe AMS. Look for

- ☐ Inability to coordinate bodily movements
- ☐ Confusion; altered mental state or consciousness
- ☐ Bluish coloration of skin or unnatural paleness
- ☐ Extreme weakness or fatigue—compare with previous state and others in field party
- ☐ Difficult or painful breathing at rest or with very mild exertion
- ☐ Greatly increased pulse rate and respiration at rest—compare with previous state and others in field party
- ☐ Rattling or bubbling sounds in chest
- ☐ Cough

Although respiratory signs and symptoms (HAPE) or neurological signs and symptoms (HACE) may be dominant, HAPE and HACE are often mixed. Group leaders should assess the condition of each member of the group every night and morning, when symptoms are likely to be most apparent.

PLANNING FOR FIELD SAFETY

If signs of severe AMS are noted, the person should descend to a lower altitude while he or she is still able to walk unassisted. Never leave a person with signs of severe AMS unattended for any length of time: the disorder can progress to more severe stages quite rapidly, particularly at night.

If severe AMS is not detected until its later stages, the victim should be considered a medical emergency. Administer oxygen, if available, and immediately transport the victim to a lower altitude—2,000 to 3,000 feet is usually sufficient. Continue oxygen treatment at the lower elevation, and allow the person to rest there for several days. Following recovery, which is usually rapid, transport or walk the person out to a medical facility for a thorough physical examination.

High altitude pulmonary edema. HAPE is caused by a reduced concentration of oxygen in the bloodstream, which may lead to death.

HAPE usually occurs when unacclimatized people ascend quickly to high altitude (above 8,000 feet) and then immediately engage in strenuous activity. Very rapid ascents by acclimatized people can also lead to the disorder. Those under 18 years old are especially susceptible to the disorder.

High altitude cerebral edema. HACE, a disorder marked by brain dysfunction, is generally confined to altitudes above about 12,000 feet, although deaths have occurred from it at lower altitudes. Most episodes of HACE occur in people who have spent several days at altitudes above 12,000 feet. The disorder is very dangerous.

HAZARDS OF SPECIFIC REGIONS AND SETTINGS

Its onset can be sudden and its progression rapid, even in well-conditioned climbers who have made every effort to acclimatize gradually to high altitude.

Any sign of confusion or muscular incoordination in combination with severe headache should suggest the possibility of cerebral edema. Those showing the signs or symptoms described earlier in this chapter should be given oxygen, if available, and immediately taken to a lower altitude. Because muscular incoordination can worsen quite rapidly, the affected person should be accompanied down the mountain, even if he or she seems able to travel alone. Following recovery, the person should climb down the mountain rather than rejoin the group.

If the victim becomes unconscious, he or she should be evacuated to a medical facility as quickly as possible.

Alpine Regions

If working in an area with snowfields and glaciers, learn the hazards and proper procedures for safe travel, and be sure to go properly equipped. Basic equipment for alpine work includes an ice ax, a climbing rope, and crampons. As with all safety equipment, this gear should be kept in good condition throughout your trip, and anyone planning to use it should receive prior training.

Streams coming from snowfields and glaciers can turn into raging torrents after a day of melting in the sun. You may therefore find an unexpected river blocking your return to camp after a day's traverse. Always make sure

PLANNING FOR FIELD SAFETY

you have enough equipment on hand for an emergency bivouac.

Keep away from ice cliffs and hanging glaciers, as there is danger of icefall, especially during the afternoon and evening.

Be on the lookout for cornices. Always try to approach them from the downwind side. Look for fractures, especially on ridge crests. Many people have been killed when the cornice they were standing on collapsed.

Avalanches are a serious danger in snow country, particularly following heavy snowfalls. Watch for chute-like clearings and scarred trees in otherwise heavy timberland; such clearings may indicate the path of periodic avalanches. Try to avoid crossing steep snowfields; you may create a fracture line across the field and trigger an avalanche. There is no sure way to tell if a given slope is stable or prone to avalanches; unstable, fractured snow could be hidden beneath new snow.

Snowfields are inviting in that they often provide easy access routes—but the extra time and physical effort needed to go around them are well worth avoiding treacherous and deadly situations. Where possible, stick to ridges above steep snowfields or hike in the valley below, staying well clear of the base of the slope.

If your group must cross a snowfield, do so one person at a time, each trailing a 50-foot length of fluorescent rope that can be traced by rescuers in the event of an avalanche. Loosen your pack so it can be discarded quickly in an emergency. Because the surface of the snowfield will vary depending on sun exposure during

HAZARDS OF SPECIFIC REGIONS AND SETTINGS

the day, try to cross when the surface is firm enough to support your weight, but not so firm as to make finding footholds difficult.

If you are caught in an avalanche, discard your pack and assume the fetal position immediately; this position will enable you to use your arms and legs under the snow to dig an air space or dig yourself out. If you become trapped with arms and legs extended, you won't be able to move.

A person buried under the snow can last no more than a couple of hours. If he or she is to survive, it is the other members of the group that must find the victim; there is no time to go for help.

Never cross a snowfield or attempt to glissade without an ice ax and the knowledge of how to do a self-arrest. Also, carefully investigate the conditions at the base of a steep snowfield, and consider the consequences of an uncontrolled slide.

Avoid crossing snow-covered glaciers as they harbor hidden crevasses. If such a crossing is necessary, do so equipped with ice axes and climbing ropes. Know how to search for crevasses. Learn the rescue procedures should a member of your party fall into a crevasse. The crevasse-free areas of last week may harbor new fractures obscured by snow.

Never build fires in alpine meadows, except in an emergency. Use a stove for cooking.

Don't dig holes or drive off established road tracks. Be careful when driving a snowmobile over ground that is covered by thin snow to avoid damaging the surface underneath. It takes hundreds of years for alpine tundra or meadows to recover from such damage.

Never travel alone.

Take precautions to avoid sunburn and snow blindness.

Tundra

Only build fires in an emergency. A tundra fire is usually uncontrollable and can be more destructive than a forest fire. Use a stove for cooking.

Growth of vegetation and recovery rates are extremely slow in tundra regions. Fire and vehicular damage often require decades to hundreds of years for vegetative recovery.

Wear inexpensive but comfortable work boots that drain and dry quickly when working on tundra; your boots will often get wet from frequent muskeg crossings. The acidity of muskeg waters can destroy stitching on boots in a few weeks, so bring extra footwear. High rubber boots are probably your best bet if you don't plan to do much climbing in rocky areas.

Alveolar hydatid, a disease transmitted by improperly cooked game, is endemic to St. Lawrence Island and western Alaska. See "Contaminated Food" for more information.

Plan traverses with the sun at your back. A nonsetting arctic sun can be merciless. Be sure to bring sunglasses, and protect yourself from harmful rays.

HAZARDS OF SPECIFIC REGIONS AND SETTINGS

Tropics

A great danger in the tropics is the marijuana (pakalolo in Hawaii) growers, who may be violently possessive about their crop. You may stumble onto a marijuana plot anywhere—such as in cane fields or up in the mountains. Try to make it obvious that you are not interested in the crop and move away quickly. Steer clear of smugglers, drug or otherwise.

Bear in mind that hot weather will exacerbate the effects of altitude; you may suffer altitude sickness at relatively low altitudes, especially if you are working hard. Usually you'll just work more slowly and get less done, but if you start showing signs of serious altitude sickness, immediately hike down to a lower altitude and allow yourself to acclimatize more gradually.

If warnings of high winds are announced on the radio, get off the mountain; you can't work in 70 mph winds, and the exposure hazard is extreme. Remember that even though you're in the tropics, very cold weather and even blizzards can hit at high altitudes. Be prepared.

Hawaii

Because much geologic fieldwork in the tropics is carried out in Hawaii, a few guidelines specific to that island chain are in order.

If you plan to work near an eruption, check with a monitoring station first to make sure the area is safe.

Fieldwork on older volcanoes can be hazardous. Most exposed rocks are deeply weathered and extremely

PLANNING FOR FIELD SAFETY

friable. Lava tubes may have thin roofs that can collapse if you walk on them. Don't try any technical rock climbing. If you are headed down a steep slope that you haven't climbed, be very careful: slopes often steepen to cliffs, and protective vegetation stops without warning.

Keep out of steep gullies in heavy rain; flash floods drown a couple of people every year.

If you are working on beaches or wave terraces, keep a weather eye on the ocean. Big surf regularly washes people away. The danger is in becoming preoccupied with your work and forgetting that the ocean is there. In Hawaii, the most treacherous places to work are terraces on south-facing shores during the summer months because big waves from storms in the Southern Ocean come in isolated sets about once every half hour. Listen to the surf report to get an idea of the danger. Winter waves on north-facing shores are more obvious; sometimes these waves can be immense. If a high surf watch or warning is issued, stop working. Needless to say, if the tsunami siren goes off, head for higher ground (note that the system is tested at 11 a.m. on the first of every month).

Wear a thick-soled pair of boots with heavy leather uppers for walking on lava or carbonate beach rock—these rocks are unexpectedly sharp and will cut feet and light shoes to shreds.

Fortunately, a lot of the land in Hawaii is federally or state owned, but always be sure you have permission to be where you are. You can gain access to most military land if you have a good reason, but be sure to apply for

HAZARDS OF SPECIFIC REGIONS AND SETTINGS

permission a few weeks before you need access. Many ranch owners are touchy about their property, and most private roads are patrolled.

For additional information on fieldwork in Hawaii, contact the USGS, Hawaiian Volcano Observatory, P.O. Box 51, Hawaii National Park, Hawaii 96718.

High and Low Desert and Semiarid Regions

Wear shoes with thick soles and leather uppers to insulate your feet from the hot ground and to protect your feet from cactus spines. Try to anticipate where blisters might form, and use moleskin or bandage strips accordingly. If blisters do form, leave them alone to prevent infection. Covering the area around a blister with moleskin will help protect it. If painful, however, lance a blister with a sterilized needle, and cover with triple antibiotic ointment and a sterile dressing.

The best season to work in some desert areas (e.g., the Great Basin of Nevada) is late spring, when rainfall is light and temperatures are not yet unbearable.

In the high desert (above 4,000 feet), the temperature during midsummer evenings can be in the 40s. Prepare accordingly. Early autumn snowfall is common in high desert regions. Saharan nighttime temperatures can easily be subfreezing during the winter months.

Ranching trails are common in the American West. Recent aerial photos and topographic maps (less than 10 years old) may show their locations.

PLANNING FOR FIELD SAFETY

In countries outside the United States, maps can be unreliable. Check with local oil companies or a fieldworker who has been in the area for correct information. Michelin travel guides can provide quite good road information, with all waterholes labeled.

Areas affected by military action may contain active land mines. Watch where you go. Don't drive at night anywhere near a minefield.

In semiarid regions, be extremely careful not to start grass fires or forest fires.

Temperate Forested Regions

Read "Cold Weather," even if you'll be working during the summer. Cool, wet weather can be hazardous.

Outcrops are sometimes hidden among the trees. It is important to be able to use your compass and to be able to find your way through the forest and the underbrush. *Be Expert with Map and Compass,* by Bjorn Kjellstrom, is an excellent reference.

Be wary of walking through logging slash. Avoid it whenever possible. Watch out for sharp stumps or branches that could impale you if you fall.

Always carry rain gear; the weather is apt to change suddenly, and late afternoon storms are common during the summer.

Many areas are now national or state parks. Make sure you get permission to work there. New laws for rock or fossil collection may be in effect, so check before you collect.

HAZARDS OF SPECIFIC REGIONS AND SETTINGS

Read the section on "Poisonous Plants."

In areas that are heavily populated, carry some form of identification. A letter of introduction or a business card may be helpful. Make sure you have permission to travel on private land.

In dense underbrush, it is a good idea to leave a trail of brightly colored flagging tape that you can follow on your way out. Be sure to remove the tape as you leave.

In the southern United States, poisonous snakes are common and pose a danger.

Many creatures will try to get at your food. If you're in bear country, read "Bears" for ways to avoid these animals. If you're in an area with porcupines, be aware that they will chew and destroy anything salty (even your sweaty hiking boots), so suspend food and boots from a tree, away from branches. Mice will always find your food unless it's in metal containers; fortunately, these animals do only minor damage. Many animals, such as chipmunks and Gray and Steller's jays, will go after food that is left out but won't go after food that is kept in plastic bags and put away. Never leave food inside packs, tents, or clothing overnight; animals will chew through these to reach the food.

Carry an insect repellent. You might also want to bring a headnet.

Don't work during the local hunting season. If you must go out, wear a fluorescent orange jacket and make lots of noise.

Temperate forested areas are usually very muddy during the late spring. Many streams that are easy to cross in the late summer will be made impassable by spring runoff. There will also be seasonal swampy areas not shown on maps. Beaver can alter water levels from those shown on maps. Consult with local residents, if possible, to learn of changes.

Avoid walking through planted fields; you can damage them. Walk around crop fields.

Underground Mines and Caves

Underground caves and inactive or abandoned underground mines are tempting sites for geologic study. However, they also pose serious potential hazards, and great caution must be exercised before and during exploration.

If your fieldwork requires that you work in underground mines, you should have training in mine safety and safety equipment. Never enter a mine unless you are sure your emergency equipment is current and in good working order.

Never go underground without leaving a detailed plan and projected return time with a responsible individual on the surface who knows how to get rescue help if necessary.

Before entering an unfamiliar underground site, try to find a miner, caver, or fellow geologist in the area who can advise you on whether such entry is safe—and legal. Obtain permission to explore privately owned areas. Be sure to wear a hard hat and carry with you proper equipment for illumination and air testing. As with

HAZARDS OF SPECIFIC REGIONS AND SETTINGS

above-ground rock study, attempt no rope climbing or spelunking unless it is absolutely necessary and you and the other members of your party are fully qualified.

Never sample any rocks until you've ascertained that the workings are safe. If old timbers are collapsed or rotten or rock has fallen from backs or raises, retreat immediately. Solid-looking timbers may in fact be completely rotten and have no strength, so examine them carefully.

If you suspect the atmosphere in the cave or mine is dangerous, leave the site immediately. Underground ventilation can change on a daily basis, and oxygen levels can drop to zero in the space of 30 feet. Oxidation of sulfide deposits will remove all free oxygen from the underground space.

Mines and caves are often homes to bats, snakes, and other wild animals; take care to avoid them.

Be advised that surface rainstorms can lead to a very rapid increase in the discharge of subsurface streams, posing a drowning threat to those underground. Such dangers are prominent in karst terrain.

Fieldwork in Other Countries

Preparations

Passports, visas, and permits. Get a valid passport. You may need six weeks to obtain one.

PLANNING FOR FIELD SAFETY

Apply for the appropriate visas months in advance. You may have to go to an embassy or consular office.

Resident visas may require that you leave your passport with authorities for weeks or months. Don't relinquish your passport unless absolutely necessary. In some instances, authorities have been known to sell foreign passports. Try to get a receipt or some type of temporary identification. Meanwhile, stay on friendly terms with local authorities, but be firm enough to get your program carried out. It's a fine line to walk, requiring patience, some political maneuvering, willpower, and some knowledge of how things "get done" within the administration of a particular country.

Well before you leave, check to see what permits are necessary for driving, firearms, radio transmission, and work in military areas or national parks, and find out what sort of liability insurance is necessary, if any. A local contact person may be helpful in providing such information, getting your equipment through customs, and making local arrangements.

Fieldworkers who use explosives must remember to declare them at international borders. Make sure there is a permit for their use.

Fieldworkers carrying sediments should label samples accurately and be aware that customs officers are suspicious of sediments that look like drugs. A country may have special requirements for shipping rock samples.

Immunizations and insurance. Start getting your shots and vaccinations months in advance. Information

HAZARDS OF SPECIFIC REGIONS AND SETTINGS

on which shots are required for your destination should be available from your local medical facility. Check your medical insurance to be sure it affords adequate coverage in countries outside the United States.

All vaccinations are not administered at the same time. Typhoid shots, for example, are given twice, one month apart (so plan ahead). A cholera vaccination is good for only six months, so you should get one just before you leave. If you'll be taking malaria pills, remember that you must start taking them at least two weeks before your trip.

Equipment. Exercise judgment and clever packing to carry as little weight as possible on international trips, and leave items at home that you can get locally. Field equipment, with the notable exceptions of compass, boots, geologic hammer, and sleeping bag, may be available in the local shops.

Obtain as many references and maps as possible before leaving the United States. These are often unavailable in foreign countries or obtainable only after cutting through much red tape. Air or Landsat photos are usually available for regions where maps are unavailable, but don't rely on overseas mail to get them to you.

On the Road

Travel with care near contested international borders. The actual position of the border is often in dispute and may not correspond to that marked on your map. The careless fieldworker can find himself or herself in the custody of a border patrol of a country he or she never meant to visit.

Familiarize yourself with the local laws. Drive with much more caution than you would at home. Roads may not be well maintained, unexpected trenches or boulders may appear without warning, and local people may not obey traffic laws to which you are accustomed. As a foreigner, you will probably be considered at fault in any accident. Running down a chicken may get you into serious trouble. Don't take it as a joke. Be extra careful in small towns and when passing buses unloading passengers. In many countries it is not advisable to drive at night.

Local Conditions

Upon arrival, inform the local authorities of the nature and expected duration of your fieldwork. It's also a good idea to check in with the U.S. Consulate and file an itinerary.

You should, of course, communicate with the local geologist or geologic survey prior to your arrival to obtain their permission and advice concerning your project. It would be foolish to duplicate the research of the local geologists.

Always remember that you are a guest in a foreign country and act accordingly. Respect their customs.

Don't attract attention by wearing shorts if it is not the custom of the country. Shirtless attire or pants on women are inappropriate in some cultures.

HAZARDS OF SPECIFIC REGIONS AND SETTINGS

Do your best to learn the local language. Even if you know just a smattering, people are friendlier and more helpful if they see that you are making an effort.

Food and water. Take your health seriously. You can be tough and brave—eat everything, drink anything, sleep anywhere—but you risk (at the very least) losing valuable time in the field. Don't drink any water unless you have treated or filtered it; just because the locals or animals drink the water doesn't mean it's safe for you. Local soft drinks may be made using unpurified water. Internationally known soft drinks and beer are usually safe. Don't drink water in restaurants or order any drink with ice in it. Don't drink unpasteurized milk or eat uncooked foods or foods such as fresh fruit and vegetables unless you know something about their history or can easily peel or otherwise clean them. These restrictions do not apply everywhere on the globe but certainly should be heeded in many countries. They apply in major cities as well as in countryside villages.

Emergency aid. Don't count on local or national authorities to provide search and rescue services. One may hope for rapid and effective cooperation by responsible authorities in case of need, but every effort must be made to be self-sufficient.

Before you head into the bush, make sure you and your partners know how and where to obtain emergency medical attention. It would be foolish to drive 20 miles to a village in an emergency only to find that the nearest physician was 40 miles in the opposite direction.

Local help. Hiring temporary field assistants should be no problem. Virtually everywhere you go you will find many people willing to work for days, weeks, or months. Try to find someone who is knowledgeable about both field matters and town problems, preferably someone with a driver's license (be sure to include the person's name on the car rental form, if possible).

Be sure to obtain permission for access to private land. It can often be a major undertaking just to get a key to open a gate. The landowners often don't live on the land and the person with the key may be in town or out in the field for the day. Rarely are there any problems once you have located the proper people. They may even let you make a copy of the gate key. Such negotiations are facilitated if you have hired good assistants.

Wild dogs. In some countries, wild dogs are commonly mistreated and are apt to be vicious. Watch out for them.

Urban Environment

Hazardous working conditions are not unique to the wilderness; the urban environment offers dangers of its own.

Be sure to obtain permission before going onto private land.

Roadcuts are particularly dangerous places to work. It is always advisable to visit the cut before a planned field trip to determine how best to approach the site and where to park vehicles. When working, keep an eye out

HAZARDS OF SPECIFIC REGIONS AND SETTINGS

for cars. Warn each other by shouting "car." If you'll be working for some time, set up reflective warning markers. On blind corners, one person should be posted to warn others about approaching cars.

Try not to knock rocks onto the road. Retrieve such rocks as quickly as safety permits.

Choose your spot for leaving vehicles carefully. Don't let a vehicle stick out into the road; on the other hand, don't drive into a drainage ditch, either.

Be careful crossing railroad tracks, even if they look abandoned. You should be at least one track's width away from a passing train.

Day Trips

Before the Trip

Obtain all necessary permissions and permits well in advance of your trip. Find out if it's okay to collect samples from the area you'll be visiting before allowing students to do so. If you'll be working on highways for extended periods, notify local police. With groups, you will often be required to sign a release form before being granted access to a mine, quarry, or other potentially dangerous geologic sites. Such forms absolve the owner of the site of any liability should a member of your group sustain an injury on the owner's property.

Make sure the sponsoring organization carries adequate liability insurance to cover field accidents. Some

PLANNING FOR FIELD SAFETY

organizations require that trip participants either show proof of adequate personal insurance coverage or purchase short-term catastrophic health insurance policies. Many organizations also require that participants sign release forms before the trip.

Make sure all participants are wearing proper shoes (no bare feet or shoes such as high heels, clogs, or sandals) and clothing adequate for the expected weather conditions.

Make sure everyone can identify poisonous plants and other hazards of the trip.

Verify that each vehicle contains a copy of its insurance coverage and a basic medical kit.

In the Field

Driving to and from field sites is probably the most dangerous activity of the day. Drivers should be alert and drive with extreme care; they are responsible for many lives. Allow no consumption of alcoholic beverages in vehicles or by drivers.

Vehicles should stick together so that no one gets lost. It may be helpful to equip each vehicle with a portable CB radio or short-range commercial FM radio, both to maintain group communication and to allow members of each vehicle to point out local geologic features to those in other vehicles.

Make sure hammers are used correctly.

If participants are working on rock slopes, make sure they don't work directly above and below one another.

HAZARDS OF SPECIFIC REGIONS AND SETTINGS

Don't allow them to climb or work on rock overhangs, particularly if others are below.

Avoid steep rock faces unless members of the group have had proper training and are appropriately equipped; dislodged rocks can be a hazard. Particularly dangerous are walls in abandoned quarries. These walls are artificially steep, and nature is in the process of trying to achieve gentler slopes. Keep everyone away from old machinery and abandoned explosives.

Pick up litter; leave the site clean.

Account for everyone after every stop. Use a buddy system to promote safety in the field and to check that no one has been left. If possible, assign each person a number before the first stop and have them count off before leaving each site. More than one person has been left behind at field stops.

8
IN CASE OF EMERGENCY

The following information is provided as an aid to those who have already completed certified courses in first aid and cardiopulmonary resuscitation. It outlines basic emergency response procedures, but is by no means complete. It should not be considered a substitute for proper medical training.

Bear in mind that if you provide medical assistance to an injured person, you can be held legally liable for any harm you cause that person through improper procedure or carelessness. The law does, however, make allowances for the extenuating circumstances surrounding a wilderness emergency.

Patient Assessment System

The fieldworker responding to an emergency has three principal goals:

1. Identify and correct life-threatening problems.
2. Identify and treat secondary injuries.
3. Stabilize the patient's condition, and continue to monitor it for signs of improvement or deterioration.

PLANNING FOR FIELD SAFETY

The patient assessment system is designed to help the rescuer achieve those goals in an orderly, effective, and efficient manner.

Primary Assessment

The purpose of the primary assessment is to ascertain quickly the nature of the accident and identify and correct any life-threatening problems.

Survey the accident scene. This initial survey should be done before or during your approach to the victim and should not require extra time.

Quickly determine the following:

Whether you can reach the victim without risking injury to yourself.

Whether the person can remain where he or she is for the time being, or if the victim must be moved to ensure his or her safety (e.g., is the person lying in the path of a potential rockslide or trapped in a burning vehicle?). The victim should be moved only if he or she is in immediate danger of death or additional injury at the accident site; until the nature and severity of the injury or injuries have been determined, the act of moving the victim may have life-threatening implications. If the victim must be moved and you suspect a spinal injury, immobilize the neck and back according to standard first aid procedure, and move the person as a unit to prevent spinal cord damage. If the victim is not in immediate danger at the accident site, the maxim "treat him where he lies" should be followed.

IN CASE OF EMERGENCY

Whether the injury is the result of trauma (e.g., a fall) or illness (e.g., heart attack). Both may be possible. Try to ascertain the cause of the injury if the accident was not witnessed. Determine whether spinal injury is a possibility.

Perform a primary survey. The purpose of the primary survey is to detect and correct any life-threatening problems. As a memory aid, these problems are referred to as the ABCD's of emergency care:

A—Airway. The airway must be open and unobstructed.

B—Breathing. The victim must be breathing effectively.

C—Circulation. The heart must be beating effectively, and there must be no profuse bleeding.

D—Disability. The status of the nervous system must be determined. Is the victim conscious or unconscious? Does the victim have a spinal injury?

Checking the ABCD's:

Check for airway obstruction. Tap the victim on the shoulder, and ask if he or she is OK. If the victim is able to speak, you know that the airway is OK. If the victim cannot speak but is conscious, or if the victim is unconscious, make sure the airway is clear. Consult your first aid manual for help. If there's a possibility of spinal injury, avoid moving the patient's neck.

Check for breathing. Use the "Look, Listen, and Feel" technique: look for chest movement; listen for air being inhaled and exhaled; and feel for air being exhaled from the nose and mouth. If the victim is not breathing, begin artificial respiration.

Check for circulation. Feel for a pulse in the carotid artery, which is on either side of the neck. If no pulse is detected, begin cardiopulmonary resuscitation.

Control profuse bleeding by applying direct pressure over the bleeding site. Consult your first aid manual for information on controlling arterial bleeding and on treating shock.

Check the status of the nervous system. If you suspect a spinal injury, refer to your first aid manual for instructions on how to protect the spinal cord.

Secondary Assessment

After life-threatening injuries have been treated, the rescuer must perform a secondary assessment to identify and treat nonlife-threatening injuries. Performing a secondary survey, monitoring the patient's vital signs, and developing a course of treatment are critical, even if the victim's injuries seem obvious. It is easy to be distracted by a fracture or other obvious injury and fail to notice a less obvious injury that may pose a more serious threat.

As you perform the secondary assessment, make sure you keep an accurate chronological record of the victim's condition, vital signs, and the treatment given. It is extremely valuable to have a prepared patient assessment form that you (or, preferably, a partner) can fill out as you go through the evaluation. See "Sample Patient Assessment Form."

Perform a secondary survey. The purpose of the secondary survey is to examine and evaluate all areas of the victim's body for injuries. Keep in mind that doing a proper physical evaluation takes practice. All field personnel should have gone through a first aid training program and should practice the secondary survey. Basic first aid skills are essential for properly evaluating the victim and initiating the necessary treatment.

If the victim is conscious, determine the nature and cause of the injury or injuries by asking yes and no questions (e.g., are you dizzy?) and open-ended questions (e.g., where do you hurt?). The following are some common questions to help you gather as much information as possible.

> Do you hurt anywhere?
> Do you remember what happened? What?
> Are you allergic to anything? What?
> Do you feel short of breath?
> Do you feel nauseated?
> Did you lose consciousness? Do you know for how long?
> Are you taking any medications? What? When did you last take medications?
> Do you feel wet anywhere? Where?
> When did you eat last? What?
> When did you drink last? What?

As a memory aid, the acronym AMPLE is helpful when questioning a victim:

A—Allergies
M—Medications
P—Past history
L—Last meal
E—Events

Examine the victim from head to toe using the senses of sight, hearing, touch, and smell. Gently walk fingers over a portion of the victim's body, then gently press palms flat against that portion to feel for any abnormality. Look and feel for signs of bleeding, pain, numbness, swelling, discoloration, or deformity. It is important to ask the conscious victim questions during the examination to find out if he or she feels anything unusual in a particular area. Make the examination of the victim's head, neck, chest, abdomen, pelvis, back, legs, and arms as comprehensive as possible. If possible, have any physical examination carried out by a person of the same sex.

Monitor vital signs. Measure and record the victim's vital signs to assess the status of the victim's circulatory, respiratory, and nervous systems. Although a single measurement is useful, it is important to periodically recheck all vital signs to monitor the victim's progress. Examine the following vital signs:

Pulse. Take at the wrists, the side of the neck, or the groin. Normal pulse rate for a healthy adult is 60 to 80 beats per minute. This rate may be significantly lower for athletes. Also note whether the pulse is regular or irregular.

Respiration. Test by listening, watching, and feeling the chest for movement. Make sure that air is being exchanged when using chest movement as your indicator. Normal respiration rate in most adults is 12 to 20 breaths per minute. Note also the character of the breathing—full, shallow, gasping, or labored.

Blood pressure. Take with a blood pressure cuff (sphygmomanometer) and stethoscope, if available. Changes in blood pressure indicate changes in the volume of circulating blood, the capacity of the blood vessels, or the ability of the heart to pump. Blood pressure can fall rapidly in cases of severe hemorrhage, heart failure, or shock.

Blood pressure changes with pulse rate. If pulse rate remains unchanged from one reading to the next, it is generally not necessary to take another blood pressure reading.

Skin color, temperature, moisture. For lightly pigmented people, establish skin color simply by looking at the skin. For darkly pigmented people, examine the fingernail beds or the mucous membranes inside the mouth or under the eyelids; these areas will show the same color changes as lightly pigmented skin.

A flushed color indicates high blood pressure or heat stroke. A cherry red color indicates carbon monoxide poisoning.

A pale, ashen color indicates insufficient circulation from shock or fright.

A bluish color, particularly in the face, indicates insufficient oxygenation of the blood. In extreme cases of carbon monoxide poisoning, the individual may have turned blue from lack of oxygen.

Check skin temperature by feeling the patient's forehead with the back of your hand. Determine if the skin feels normal, warm, hot, cool, or cold. Also note skin moisture; that is, whether the skin feels dry, moist, or clammy. The temperature and moisture of the skin can alert you to such underlying problems as inadequate circulation to certain parts of the body, chills, fever, heat stroke, inflammation, infection, and other conditions. Periodically check skin condition as you perform your patient assessment, and be alert to changes that may indicate improvement or deterioration in the underlying conditions.

Body temperature. Take with a thermometer orally or axially (in the armpit). It is best to avoid taking rectal temperature readings unless you've had previous experience. Note that temperature taken orally is always lower than temperature taken rectally. Be aware that traditional "normal" body temperature, 98.6°F (37.0°C), is simply an average; typically, normal body temperature ranges a degree or two above or below this figure throughout the day.

Level of consciousness. The patient's level of consciousness is the single most reliable way to assess the status of the central nervous system, especially in cases of head injuries. It is commonly measured on a four-point scale known as the AVPU scale. A decrease in level of consciousness generally indicates a worsening condition.

A—conscious and Alert; aware of self and surroundings.

V—responds to Verbal stimulus, but is sluggish and may be confused about general information, such as location and time.

P—not conscious but will respond to Painful stimulus by withdrawing (induce pain by squeezing earlobes or rubbing knuckles on sternum).

U—Unresponsive (unconscious); will not respond to pain.

Develop a course of treatment. Use the SOAP acronym to help you organize the patient information you've collected into an effective course of treatment. The acronym stands for Subjective, Objective, Assessment, and Plan, and is applied as follows:

Based on *subjective* information provided by the victim (e.g., medical history and areas of pain) and *objective* information obtained during the secondary survey, *assess* the victim's overall condition and develop a list of potential problems. These problems may be general (e.g., fever, vomiting, abdominal pain) or specific (e.g., appendicitis).

From the problem list, develop a treatment *plan*. If multiple problems exist, list the treatment for each problem, and assemble them into a single treatment plan.

A patient assessment form based on the SOAP acronym is included at the end of this chapter. Follow the sample form as you assess and record the patient's history and condition to help you arrive at an effective course of treatment. Your assessment will also provide a written record of the patient's condition, which may be vital to health professionals if the patient is evacuated to a medical facility.

Before beginning any treatment, stop and think. Take time to get organized. Remember that in many situations evacuation of the victim is preferable to treatment in the field.

Evacuation Procedures

Evacuating an injured person can be extremely difficult. If approached incorrectly, it can exacerbate the victim's injuries and place the rescuers at risk.

Before attempting to evacuate someone yourself, consider possible alternatives. Does the person really need to be evacuated? Perhaps, as in the case of a badly sprained ankle or other minor injury, a few days' rest at base camp is all that's needed.

If the person must be evacuated, is professional help available? It's important to find out before you head into the field what rescue resources—Park Service, Forest Service, local police—are available in your area; the types of emergencies they respond to; and how to contact them in an emergency. Can the injured person be left where he or she is while you or someone in your group goes for help? If someone goes for help, that person should carry a written record of the nature of the accident, the exact location of the victim (marked on a map), the condition of the victim (use the patient assessment form), and the type of help needed.

Remember your prearranged check-in system. If your group doesn't return to camp soon, will a search party be sent out? Perhaps it makes more sense to stay with the victim and wait for help to arrive.

If you decide to evacuate, practice and perfect any methods of moving the victim (raising, lowering, and transporting) with an uninjured person before attempting to move the victim. You should plan all aspects of the evacuation route before setting out. Getting lost or finding yourself unable to transport the victim across a river or other unexpected obstacle will waste precious hours. Inform the victim of how the evacuation will proceed and keep him or her apprised of what you are doing at all times. The victim may already be agitated, and it is up to you to keep him or her calm and reassured. The rescuers should also perform the following tasks:

Keep the victim warm. A hat can help the victim retain body heat.

During the day, protect the victim's face (especially the eyes) from the sun, using a hat, sunscreen, and sunglasses.

Keep a chronological record of the victim's vital signs and of all first aid procedures used during the evacuation.

An excellent resource for information on wilderness evacuation techniques is *Wilderness Search and Rescue* by Tim Setnicka.

Sample Patient Assessment Form

The patient assessment form is designed to help rescuers handle an emergency situation quickly and effectively. Filling out such a form as you assess the situation and the patient's condition provides an accurate record of the accident location, the victim's medical history and vital signs, and all treatment given. Such information can be of critical importance when tracking a victim's progress in the field, sending a group member for outside help, or transferring the victim to the care of a medical professional.

Ideally, one person should perform the patient assessment while another records relevant information on the form.

SAMPLE PATIENT ASSESSMENT FORM

Patient Name_____ Date _____

Location_____

Subjective Information

History

Symptoms

IN CASE OF EMERGENCY

Objective Information

Examination

Head
Eyes
Ears
Nose
Mouth/throat
Neck
Chest
Abdomen
Pelvis
Back
Legs
Arms

Vital Signs (monitor periodically)

Pulse

Respiration

Blood pressure

Skin color/temperature/moisture

Body temperature

Level of consciousness

Assessment

Treatment Plan

9

PRECAUTIONS FOR THE GROUP LEADER

This guide has focused primarily on preparing the individual for work in the field. It has emphasized the responsibility each person has to prepare for and maintain his or her own safety. On group trips, the primary responsibility for safety still lies with the individual; however, a measure of this responsibility is transferred to the group leader. It is essential, therefore, that group leaders be acutely aware of safety practices and bear in mind specific preparations and precautions peculiar to the group setting. This chapter, consisting largely of information drawn from other parts of this guide, is designed to help the group leader prepare for his or her important role.

Become certified in first aid and cardiopulmonary resuscitation. If you were certified years ago, take a refresher course. Contact your local American Red Cross chapter for course information.

Read this guide and an up-to-date first aid book. Make safety the primary consideration as you plan all aspects of your trip. Discuss safety matters—such as medical kit size and organization, evacuation procedures, rules for

PLANNING FOR FIELD SAFETY

tool use—with the other group leaders, and establish a list of specific safety rules to be followed by all members of the group.

Require each person to fill out a confidential medical history form that includes current physical condition; current state of health; allergies to drugs, insects, or other substances; dietary restrictions; high or low blood sugar; reactions to temperature extremes; reactions to altitude; high or low blood pressure; heart conditions; physical handicaps; and diabetes. Keep copies of these forms in the base camp medical kit.

If a group member needs any prescription medication, make sure he or she carries an adequate supply. It's a good idea to keep a back-up supply in the base camp medical kit.

For each member of the field party, obtain the name, address, and phone number of a person to contact in case of emergency. Compile this information into a list, and make sure that at least one person—preferably a group leader—in each vehicle has a copy. It's also a good idea to include the names and phone numbers of officials from the organization that is sponsoring the trip. You might need to contact them in an emergency.

If you're heading out for longer than a day, make sure students notify their parents or guardians of where they're going and what they'll be doing.

Make sure your school or other sponsoring organization carries adequate liability insurance to cover field accidents. Some organizations require that trip participants either show proof of adequate personal insurance

PRECAUTIONS FOR THE GROUP LEADER

coverage or purchase short-term catastrophic health insurance policies.

Make sure each group leader has a large-scale topographic map of the trip area and that his or her orienteering skills are good.

Before heading into the field, assemble the group and discuss the plans and goals of the upcoming trip. Stress safety. Lay down all ground rules to be followed by group members. Emphasize repeatedly the importance of common sense, good judgment, alertness, and adherence to safety rules. Review basic first aid, and outline evacuation procedures. Discuss the unique hazards of the specific region you'll be working in and how to deal with them. Ideally, group members should be required to read this guide.

Demonstrate the proper use of radios, stoves, backpacks, tents, tools, and other gear. Allow group members to practice using any equipment they'll be working with in the field. Describe the amount and type of gear each group member should carry. Set aside a time well before the trip to examine each group member's personal gear to make sure it's suited to the type of environment you expect to encounter and the work you expect to do.

Divide the main group into smaller groups, and before setting out, allow them time to practice their assigned work, such as setting up tents or cooking.

Check and double-check your equipment to make sure you haven't forgotten anything. Remember to bring all essential teaching aids.

PLANNING FOR FIELD SAFETY

Once in the field, keep an eye out for hazards, and point them out to group members before the hazards become a problem. Constantly remind group members of the need for good judgment and safety awareness. Do not force a group member to do an activity that he or she is afraid of doing. Periodically assess the physical condition of each person to make sure no one is suffering from fatigue, cold, dehydration, or other common hazards. As group leader, you must learn to think for others as well as for yourself.

Review "Day Trips."

10

SHIPPING ROCK SAMPLES

Although not exactly a matter of safety, figuring out how best to ship rock samples is nonetheless a thorn in the side of many fieldworkers. This chapter offers a few suggestions on how to make this task easier.

If you expect to be shipping rock samples back from the field, determine what methods you will use and what equipment you will need before heading out.

Be sure to carry a waterproof marking pen for labeling samples and boxes and a roll of fiberglass strapping tape for sealing and binding packages.

Rocks should be wrapped to avoid obliterating surface features or labeling. Old newspapers are probably the cheapest globally available packing material. Scrap, which is unused newsprint, is often available from newspaper offices or moving firms. Other packing materials include paper towels and sawdust.

Rocks sent back from the field should be labeled "Specimens of rock for scientific study—no commercial value" to avoid delays with customs.

In North America, truck freight companies generally provide the least expensive form of shipping and will deliver directly to your lab or university loading dock.

PLANNING FOR FIELD SAFETY

Containers

Paint cans. Use 5-gallon metal (not plastic) cans; these are available from paint stores or paint contractors. The cans come with covers and can also serve as seats or vermin-proof food storage containers.

Wooden crates. These can be built by a crating company or a local lumber yard. For ease of shipping, keep the crates relatively small. Make handles by drilling two holes in each end of a crate and looping lengths of rope through them.

Cardboard boxes. If you must use cardboard boxes, be careful not to overload them. Use relatively small boxes, and pack them carefully, making sure no sharp edges are exposed that could poke through the box walls. Use plenty of strapping tape when sealing the boxes. It's a good idea to pack your rocks in heavy burlap sacks before loading them into the boxes. If the boxes burst open, the rocks will not be lost.

Shipping from Other Countries

Before you pack your rocks, get information about special shipping requirements and options.

After carefully packing your rocks and labeling all boxes with your name and the address of their final destination, bring your shipment to the post office or the loading dock of the air freight company or shipping line you're using. Address the consignment to the import handler of the domestic shipping company that will transport the rocks once they reach the United States.

SHIPPING ROCK SAMPLES

Obviously, it's important to have this information with you before you leave the United States.

Send all relevant papers, especially the original bill of lading, on ahead to the domestic carrier. Retain a copy for yourself. Send the receipt number, the shipping date, and the domestic carrier's name to the person responsible for such shipments at the shipment's final destination, and notify this person that the shipment is on its way.

If your samples are important, be sure to send them air freight, not ocean freight. Although air freight is more expensive, it is also more reliable, and you'll end up spending a lot more money if you have to go back and collect your samples again. If using a postal service, consider sending your samples as insured or registered mail. If transporting your samples on a passenger aircraft, consider carrying high-priority samples as carry-on rather than checked baggage. Check with your airline about the cost of bringing your samples back as excess baggage. This method is sometimes an inexpensive way to ship rocks.

INFORMATION SOURCES

Advanced First Aid and Emergency Care. 2nd edition. American National Red Cross, Garden City, NY, 1979.

Angier, Bradford. *How to Stay Alive in the Woods.* Collier Books, New York, 1984.

A Pocket Guide to Cold Water Survival. U.S. Coast Guard (CG 473), 1980.

Atmospheric Environment Service, 4905 Dufferin Street, Downsview, Ontario, M3H 5T4, Canada. Contact the Assistant Deputy Minister for weather information in Canada.

Auerbach, Paul S. *Medicine for the Outdoors: A Guide to Emergency Medical Procedures and First Aid.* Little, Brown and Co., Boston, 1991.

Centers for Disease Control, 1600 Clifton Road, Atlanta, Georgia 30333; 404/639-3534.

Comparative Climatic Data. U.S. National Climatic Center, Federal Building, Ashville, NC 28801-2696.

Drilling Safety Guide. International Drilling Federation, Columbia, SC, 1991.

Environmental Protection Agency, 401 M Street, S.W., Washington, D.C. 20460; 202/260-2090.

Fletcher, Colin. *The Complete Walker III*. Alfred A. Knopf, New York, 1986.

Forgey, William W., M.D. *Wilderness Medicine*. Indiana Camp Supply Books, Pittsboro, IN, 1987.

Health Information for International Travel. U.S. Public Health Service, Superintendent of Documents, U.S. Government Printing Office, Washington, D.C. 20402.

Herrero, Stephen. *Bear Attacks: Their Causes and Avoidance*. Lyons and Burford, Publishers, New York, 1985.

Hubbell, Sue. "A killer bee by any other name...." *Smithsonian* 22 (September 1991): 116-126.

Isaac, Jeff, PA-C, and Goth, Peter, M.D. *The Outward Bound Wilderness First-Aid Handbook*. Lyons and Burford, Publishers, New York, 1991.

Kaniut, Larry. *Alaska Bear Tales*. 7th edition. Alaska Northwest Publishing Co., Anchorage, 1986.

Kjellstrom, Bjorn. *Be Expert with Map and Compass*. Charles Scribner's Sons, New York, 1976.

Manning, Harvey. *Backpacking: One Step at a Time*. 4th edition. Vintage Books, New York, 1986.

National Association for Search and Rescue Wilderness Medicine Programs, Wilderness Medical Associates, RFD 2 Box 890, Bryant Pond, ME 04219; 207/665-2707.

INFORMATION SOURCES

National Oceanic and Atmospheric Administration, National Weather Service, 8060 13th Street, Silver Spring, MD 20910; Attention: W112. Write for a list of NOAA weather stations.

Occupational Safety and Health Administration. U.S. Department of Labor, 200 Constitution Avenue, N.W., Washington, D.C. 20210; 202/523-8063.

Peters, Ed, ed. *Mountaineering: The Freedom of the Hills.* 4th edition. The Mountaineers, Seattle, 1982.

Randall, Glenn. *The Outward Bound Map and Compass Handbook.* Lyons and Burford, Publishers, New York, 1989.

Schuh, Dwight R. *Modern Outdoor Survival: Wilderness Skills and Techniques for All Conditions.* Menasha Ridge Press, Birmingham, 1989.

Setnicka, Tim J. *Wilderness Search and Rescue.* Appalachian Mountain Club Books, Boston, 1981.

Standard First Aid and Personal Safety. American National Red Cross, Garden City, NY, 1979.

Wilkerson, James A., ed. *Medicine for Mountaineering.* 3rd edition. The Mountaineers, Seattle, 1985.

INDEX

acute mountain sickness, 141-147, 151
aircraft, 69-79, 120-121
  landing, 41, 73-74, 76-79
alcohol stoves, 24, 26
allergic reactions to insects, 134-135
alligators, 126
alpine regions, 147-150
altitude sickness, 141-147, 151
  prevention, 142-143
  symptoms, 144-145, 147
  treatment, 144-147
alveolar hydatid, 50-51, 53, 150
AMPLE, 171-172
anaphylaxis, 134-135
animals, 115-127, 155-157
  bears, 115-125, 155
  cougars, 126-127
  deer, 126
  dogs, 38, 125, 126, 162
  elk, 126
  moose, 126
  porcupines, 83, 155
ANSI, 30-31
antivenin, 129-131, 136
ants, 134

Atmospheric Environment Service, 2
automobiles, 59-65
avalanches, 148-149
axes, 22, 41
ice, 147, 149

backpacks, 6, 46, 148-149
beaches, 152
bears, 115-125, 155
  avoiding, 117-122
  encounters, 122-125
  range and habits, 115-122
bees, 134-135
bilharziasis, 52
birds, 126, 155
black bears, *see* bears
black widow spiders, 135-136
boating, 79-87
brown recluse spiders, 135-136
butane stoves, 24, 26

camping, 3-8, 54-57, 121-122
  defensively, 56, 121-122
  equipment, 3-8
  fires, 8, 56-57, 80
  sites, 54-56, 83
canoes, 82-83
caves, 44, 156-157

193

Centers for Disease Control, 9, 133
chainsaws, 22-23
chiggers, 139
chisels, 21
clothing, 4, 40, 77
    around drill rigs, 31-32
    around water, 77, 80, 86-87
    cold weather, 89-93
    hot weather, 104
    hunting season, 42, 155
    materials, 90-92
cold-water immersion, 80-81, 101-102
    survival suits, 77, 80
cold weather, 89-103
    clothing, 90-93
    hazards, 93-103
compass, 23-24, 154, 159
contamination, 50-54, 161
    food, 50-51
    water, 50-54
coral snakes, 127-130
cougars, 126-127
crash landings, 74, 76
crocodiles, 126

day trips, 163-165
deer, 126
dehydration, 105-106, 143
dengue fever, 133
desert regions, 153-154
distress signals, 40, 48-49, 74, 77, 80
dogs, 38, 125, 126, 162

drilling rigs and sites, 28-35
driving, 59-60, 163-164
    defensively, 63-64
    international, 159-160
    off-road, 64-65, 149-150
dysentery, 53

elk, 126
emergency
    assessment for injuries, 167-179
    contact person, 182
    preparation for, 17, 39-41, 161
    signals, 49, 74
    supplies, 80, 148
encephalitis, 133
environmental regulations, 30
equipment, 3-8, 21-33, 39-42, 60-62, 156
evacuation plans, 18, 176-177
eye protection, 19, 42
    safety glasses, 21, 23, 31
    sunglasses, 87, 103, 104, 150

falling, 44-47, 101-102
fire, 56-58
    campfires, 56-57
    distress, 49, 80
    wildfires, 57-58, 149, 150, 154
firearms, 27-28, 71, 79, 118-119, 121
fire starters, 8

flares, 40, 49, 77
flies, 132-133
flight guidelines, 69-74
food, 49-51
   contamination, 50-51, 161
   quantities and types, 15-16, 49
   high altitude diet, 143-144
   storage, 55, 120-122, 155
footgear, 4-5, 46, 159, 164
   in desert regions, 153
   in tundra regions, 150
   on a ship, 85
   on lava, 152
   safety, 30-31
frostbite, 98-101
   prevention, 98-99
   treatment, 99-101

garbage disposal, 55, 122
gas lanterns, 26-27
gas stoves, 24-26
giardiasis, 52
glaciers, 47, 147-149
gloves, 5, 31
grizzly bears, 115-125
group leader precautions, 10, 16-18, 43, 145, 163-165, 181-184

HAPE and HACE, 141-147
hard hats, 30, 42, 156
hatchets, 22
Hawaii, 151-153
head gear, 30, 42, 87, 104, 156
hearing protection, 31

heat illness, 105-109
   prevention, 105-106
   symptoms, 107-108
   treatment, 107, 109
helicopters, 41, 74-79, 120-121
HELP, 101
hepatitis, 52
high altitude, 141-147, 151
   acclimatization, 142-143, 151
   sickness, 141-147
horses, 66-69
hot weather, 103-110, 151
   hazards, 105-110
hunting license, 119
hyperthermia, 106-109
hypothermia, 72, 93-98
   prevention, 93, 98
   symptoms, 94-95
   treatment, 95-98

ice, 47, 81
immunizations, 9, 133, 158-159
inflatable boats, 82-83
insect
   bites and stings, 132-135, 137-139
   repellent, 119, 138, 155
insurance, 59, 158-159, 163-164, 182
International Drilling Federation, 28
international fieldwork, 157-162

kerosene stoves, 24, 26
knives, 22

land mines, 154
lanterns, 26-27
latrines, 55-56
lightning, 54-55, 111-113
logbook, 42
logistics, 16-18, 42
Lyme disease, 137, 139

machetes, 22, 41
malaria, 132-133
maps, 42, 69, 153-154, 159
    waterproofing, 24, 80
marijuana plots, 39, 151
medical, 9-16, 167-179
    emergency, 167-179
    history, 9-10, 16
    kits, 10-15
medications, 9-15, 85-87, 132-137, 144
mice, 155
mines, 156-157
moose, 126
moonshiners, 39
mosquitoes, 132-133
motorboats, 81-82

National Weather Service, 2
navigation, 80

ocean research vessels, 84-87
OSHA regulations, 20, 30, 34

pack animals, 66-69
packs, *see* backpacks
passport, 157-158
patient assessment system, 167-176, 178-179
permissions, 54, 163
    private lands, 17-18, 155, 156, 162
    public lands, 154
pit vipers, 127, 129-131
poisonous plants, 139-140, 154, 164
polar bears, *see* bears
porcupines, 83, 155

quicksand, 47-48

rabies, 38, 51
radios, 1-2, 38, 74, 120, 164
roadcut cautions, 162-163
rock hammers, 21-22, 159, 164
rocks, shipping, 185-187
Rocky Mountain Spotted Fever, 137, 139
rope, 6, 41, 46, 148

safety regulations, 20, 30-34
saws, 22-23, 41
schistosomiasis, 52
scorpions, 136-137
seasickness, 86
semiarid regions, 153-154
signaling devices, 40, 48-49, 74, 77
smoke bombs, 49, 77

snakes, 127-131, 155, 157
  habits, 127-129
  snakebites, 129-131
snow, 102-103, 147-150
  blindness, 103, 150
  snowfields, 147-150
  vehicles, 65-66, 149
SOAP, 175-176
spiders, 135-137
stoves, 8, 24-26, 150
streams, 45-47, 147-148, 155-156
  crossing, 45-47, 147-148, 155-156
  subsurface, 157
sunburn, 102-103, 109-110, 150
surf cautions, 152

tarantulas, 135-136
temperate forested regions, 154-156
tents, 7-8, 122
thunderstorms, 110-113
ticks, 137-139
tidal areas, 83-84, 152
tools, 5-6, 21-23, 32-33, 41, 147, 149
trichinosis, 50
tropics, 151-153
traverses, 17, 73, 150
tsunami, 152
tundra, 150

urban fieldwork, 162-164

U.S. National Climatic Center, 2

vaccinations, *see* immunizations
vehicles, 59-66
visas, 157-158
vital signs, 172-175
volcanoes, 151-153

walking stick, 46
wasps, 134-135
water, 50-54
  contamination, 50-53
  drinking, 41, 161
  safety, 79-81, 101-102
  treatment, 53-54
weather, 89-113
  around water, 80-81
  cold, 89-103
  hazards, 93-113
  hot, 103-110
  information, 1-2
white-gas stoves, 24-25
wildfires, 57-58